AGRICULTURAL WATER MANAGEMENT

Proceedings of a Symposium on Agricultural Water Management
Arnhem / Netherlands / 18-21 June 1985

AGRICULTURAL
WATER MANAGEMENT

Edited by
A.L.M.VAN WIJK & J.WESSELING
Institute for Land and Water Management Research, Wageningen, Netherlands

Sponsored by the Commission of the European Communities,
Directorate-General for Agriculture, Coordination of Agricultural Research,
as part of the Land and Water Use and Management Programme

A.A.BALKEMA / ROTTERDAM / BOSTON / 1986

CIP-DATA KONINKLIJKE BIBLIOTHEEK, DEN HAAG

Agricultural

Agricultural water management: proceedings of a symposium on Agricultural Water Management Arnhem, Netherlands, 18-21 June 1985 / ed. by A.L.M. van Wijk & J. Wesseling.
— Rotterdam (etc.): Balkema. — Ill.
Sponsored by the Commission of the European Communities, Directorate-General for Agriculture, Coordination of Agricultural Research, as part of the Land and Water Use and Management Programme.
ISBN 90-6191-639-9 bound
SISO 631.2 UDC 626.8
Trefw.: water management; agriculture.

Publication arranged by: Commission of the European Communities,
Directorate-General Information Market & Innovation, Luxembourg

LEGAL NOTICE
Neither the Commission of the European Communities, nor any person acting on behalf of the Commission or the Ministry is responsible for the use which might be made of the following information.

EUR 10055 EN

ISBN 90 6191 639 9

Published by A.A.Balkema, P.O.Box 1675, 3000 BR Rotterdam, Netherlands
Distributed in USA & Canada by A.A.Balkema Publishers, P.O.Box 230, Accord, MA 02018

Printed in the Netherlands

Table of contents

Session 3. *Installation and maintenance of drainage systems*

Session 4. *Regional and local water management systems*

Session 5. *Effects of agriculture on its environment*

Preface

In the framework of the Land Use and Rural Resources Programme of the
Commission of the European Communities problems of water management have
been dealt with from an early stage. The attention then was mainly rocussed
on drainage problems in the so-called disadvantaged areas. The proceedings
of the first Seminar on Land Drainage, organised in Cambridge (Gardiner,
1982) contain many contributions on drainage and reclamation of soils under
difficult conditions (low permeability, seepage). Other contributions
treated the effects of drainage on crop yields and farm income and the
development of new drainage techniques. The seminar also offered a good
opportunity to discuss future research and development needs on the basis
of a presentation of each country's position. In a subsequent workshop in
Brussels in 1982 these discussions were continued and widened to subjects
that got little attention in the Cambridge seminar, such as irrigation and
environmental effects. In the meantime Greece had joined in the programme.

The (unpublished) results of the Brussels workshop were accepted by
the Programme Committee on Land and Water Use and Management. They were
very useful in listing the fields of research that could be considered as
important for many countries of the EC. The programme of the next EC-work-
shop on Agricultural Water Management, organised by the Institute for Land
and Water Management Research (ICW) from 18-21 June 1985, was drafted on
the basis of agreement on priorities reached in the preceding meetings.

Participants of the workshop acknowledged the fact that the EC made
this workshop possible and provided the means to publish the results. Since
the available budgets for agricultural research are declining in almost
every country, it is of greatest importance to promote cooperation between
institutes and scientists in different countries to make maximum use of the
available expertise. Keeping this in mind, the workshop, after drafting its
conclusions and recommendations on each of the five subjects, also put forward a number of recommendations on future EC-activities in this field.

G.A. Oosterbaan
Institute for Land and Water
Management Research

Session 1
Drainage and reclamation of soils with low permeability

Chairman: G.SPOOR
National College of Agricultural Engineering,
Silsoe, Bedford, UK

Characterization of flow processes during drainage in some Dutch heavy clay soils

J.BOUMA
Soil Survey Institute, Wageningen, Netherlands

ABSTRACT

Combined micromorphological and soil-physical studies of water movement in saturated riverine and marine heavy clay soils in the Netherlands have demonstrated the occurrence of biporous flow: relatively rapid movement of water along soil cracks and slow movement through the fine-porous natural aggregates ('peds'). These flow processes present measurement problems for K_{sat} and water table levels. Use of the column, cube and drain-cube method for measuring K_{sat} will be discussed.

Results obtained with the crust test for measuring K_{unsat} near saturation will be shown to demonstrate the sharp drop of K upon desaturation of the soil. This, in turn, results in very low drainage rates.

Tile drainage of both types of heavy clay soils resulted in a significant increase of K_{sat}. The increase in the young riverine soil was due to increased cracking, while formation of vertical worm channels caused a very high increase of K_{sat} in the older marine soil.

1. FLOW PATTERNS IN CLAY SOILS

Many heavy clay soils have relatively large cracks or root and animal burrows in a fine-porous soil matrix. This matrix has a very low hydraulic conductivity (K) and significant fluxes of water and solute through the entire soil are therefore only possible when continuous large pores are present. These large pores are unstable as their dimensions change upon swelling and shrinking of the soil following wetting and drying.

Flow of water and solutes in heavy clay soils is quite different from flow in more sandy soils in which most soil pores contribute to water movement. Flow patterns in heavy clay soils are difficult to characterize with physical methods. Measurement of fluxes provides, of course, no clues.

Breakthrough curves provide information on large-pore continuity by their point of initial breakthrough (e.g. Wösten and Bouma, 1979), but they do not indicate the functioning of various pores in the soil. It is sometimes important to know whether rapid breakthrough is due to flow along one large continuous pore or to flow along several pores. Also, flow along planar voids (cracks) has different dynamics than flow along root and worm channels. Obviously, physical methods cannot be used to distinguish different types of pores.

Morphological staining techniques have been used successfully to define

3

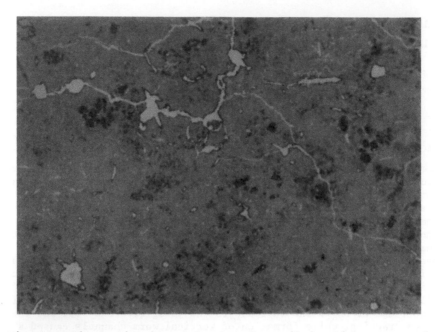

Fig. 1   Thin-section image of a wet clay soil in which water conducting
macrovoids are stained with methylene blue (dark walls of voids). Only
continuous voids are stained

flow patterns in heavy clay soils. Undisturbed samples of a clay soil that
had been close to saturation for a period of several months, were percolated with a 0.1% solution of methylene blue. The samples were freeze-dried
and thin sections were made in which stained pore walls were observed next
to morphologically identical pores with unstained walls (Figure 1). Staining indicates pore continuity which is more crucial to hydraulic conductivity than the often used pore size distribution.

The studies, cited above, resulted in the following conclusions: 1) K_{sat}
was governed by small pore necks in the flow system with diameters of
approximately 30 μm. Small changes in pore-neck sizes had a large effect on
K_{sat}. For example, a pore neck of 22 μm resulted in a K_{sat} of 5 cm·d^{-1} and a
neck of 30 μm in a K_{sat} of 25 cm·d^{-1}; 2) water-conducting (stained) larger
pores usually occupied a volume that was lower than 1%. Such pores should
therefore be characterized in terms of numbers per unit area rather than in
terms of relative volume; 3) flow occurred mainly along planar voids
(cracks). This contradicted the common assumption for Dutch clay soils that
cracks close completely upon swelling; 4) using morphological data, K_{sat} of
six different clay soils could be calculated according to:

4

$$K_{sat} = \frac{\rho g}{\eta S} \left(\frac{d_n^3 1}{12} + \frac{r_n^4}{8} \right)$$

in which ρ = liquid density (kg·m^{-3}), g = acceleration of gravity (μm·s^{-2}), η = viscosity (kg·m^{-1}·s^{-1}), S = cross-sectional area of soil (m^2) containing a length of 1 (m) of stained plane slits with neck width d_n (m) and n channels with neck radius r_n (m). Neck widths were calculated from the size distribution of stained pores, using a newly developed pore-continuity model (Bouma et al., 1979).

2. MEASURING HYDRAULIC CHARACTERISTICS

2.1. Hydraulic conductivity of saturated soil (K_{sat})

Measurement of K_{sat} in clayey soils presents the following problems: i) smearing of the walls of bore holes may yield unrealistically low K_{sat} values for the auger hole method, which are in any case an undefined mixture of K_{sat} (hor) and K_{sat} (vert); ii) small samples give poor results because of unrepresentative large-pore continuity patterns (e.g. Bouma, 1981a) and iii) water movement occurs only along some pores which occupy less than 1% by volume. These pores can be easily disturbed by compaction which may occur when sampling cylinders are pushed into the soil.

The cube method (Bouma and Dekker, 1981) avoids these problems and uses a cube of soil (25 cm x 25 cm x 25 cm) which is carved out in situ and encased in gypsum on four vertical walls (Figure 2). First, the K_{sat} (vert) is measured by determining the flux leaving the cube while a shallow head is maintained on top. Next, the cube is turned 90°. The open surfaces are closed with gypsum and the new upper and lower surfaces are exposed. Again, K_{sat} is measured which now represents K_{sat} (hor) of the soil in situ.

The column method measures only K_{sat} (vert) because it uses a cylindrical column of soil with an infiltrometer on top (Bouma, 1977). A particular version of this method was developed to measure K_{sat} above and below tile drains (Bouma et al., 1981). The technique involves excavation of a cube of soil (25 cm x 25 cm x 25 cm) in situ around a tile fragment. Cutting of the tile leaves it protruding from the cube for about 5 cm on both sides, also when covering five sides of the cube with a layer of gypsum outside the pit. Flow rates from the tile are measured for two conditions: 1) infiltration into the layer which was previously the lower horizontal soil surface of the cube, and 2) infiltration into the upper surface. The cube is turned upside down to allow the first measurement. The open surface is then covered with

Fig. 2   The cube method for measuring the vertical and horizontal K_{sat} of soils with macropores, that require undisturbed, large samples. The shown position will measure the K_{sat}-horizontal. The vertical face, visible on the right side of the picture, was the original upper horizontal face of the cube in the soil pit

gypsum. The cube is turned upside down again and gypsum is removed from the upper surface of the cube (which is the original upper surface of the cube in the pit) to allow the second measurement. The first measurement mainly characterizes undisturbed soil below the drain, the second characterizes soil above the drain (fill). Dye is added to the percolating water to trace patterns of water movement, using micromorphometric techniques. Flow rates are transformed to K_{sat}-values by using an electrolytic analog model.

2.2. Hydraulic conductivity of unsaturated soil (K_{unsat})

Attention will be confined here to K_{unsat} near saturation, which is particularly relevant for soil drainage. Non steady state methods, which are widely used to measure K_{unsat}, are not suitable to obtain K-values near saturation in all soils, in the range h = 0 cm to, say, h = -15 cm. These values are particularly relevant for describing water flow in clay soils. In

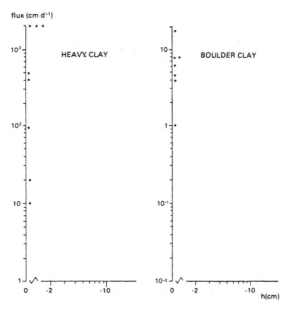

Fig. 3   K-curves of a heavy clay and a boulder clay that show the strong drop of K upon desaturation of the soil. The drop is due to emptying of macropores (after Bouma, 1982)

these soils there is a strong drop of K upon desaturation owing to emptying of the macropores (Figure 3) (see also Bouma, 1982). Again, large samples are needed to obtain representative results.

The cube method can be extended to provide K_{unsat} data near saturation. This procedure represents a version of the crust test (Bouma et al., 1983). Two tensiometers are placed about 2 and 4 cm below the surface of infiltration, which is covered by a series of crusts, composed of mixtures of sand and quick-setting cement. Earlier, the crust test used gypsum but this may dissolve too rapidly. Dry sand and cement are thoroughly mixed. Water is applied and a paste is formed which is applied as a 0.5 cm to 1 cm thick crust on top of the cube. The crust, which has perfect contact with the underlying soil because of the application method, hardens within 15 minutes. Light crusts (5 to 10% of cement by volume) induce pressure heads (h) near saturation and relatively high fluxes. Heavier crusts (20% cement and more) induce lower h-values and fluxes. Cement can be added to existing crusts to avoid removal of crusts which could cause damage. Fluxes, when steady, are equal to K_{unsat} at the measured subcrust h-value. Cubes can be placed on a sandbox to create a semi-infinite porous medium. Thus, the range of fluxes can be extended to corresponding pressure heads of approx. -60 cm.

2.3. Water table levels

Measurement of water table levels in clay soils may offer particular problems as will be illustrated by the following field study that was made in a heavy Dutch clay soil (very fine, mixed, mesic Typic Fluvaquents, according to Soil Survey Staff, 1975). The soil had the following horizons: A1g: 0-8 cm, dark grayish brown mottled clay, moderate medium compounds prisms parting to strong fine subangular blocky peds; B1g: 8-25 cm, dark gray mottled clay; B21g: 25-50 cm, dark gray mottled clay, strong coarse compound prisms parting to strong, medium angular blocky peds; A11b: 50-64 cm, dark gray clay (old surface layer); B22bg: 64-94 cm, dark gray clay, strong very coarse smooth prisms; A12b: 94-106 cm, buried surface horizon; B23bg: 106-120 cm, gray mottled clay with a compound prismatic structure parting to strong fine angular blocky peds.

Water tables were observed using: 1) a 1.2 m deep perforated pipe, 2) two series of shallow, unlined boreholes to depths of 30 and 50 cm below the surface, and 3) one series of piezometers to 30 cm. All holes had a diameter of 2 cm, as required in clay soils. Pressure heads were measured in duplicate by transducer tensiometry at 10 cm depth intervals. Different water levels were observed in the unlined boreholes at the shallow depths, while pressure heads in the surrounding soil were negative. Only the level in the deep hole corresponded with the level of zero pressure head, as obtained by tensiometry.

Occurrence of water in the shallow holes, which initially seemed to suggest the presence of a shallow water table, was explained by applying artificial rain containing chloride as a tracer. An estimated 10 percent of this rain was absorbed by the upper 20 cm of the soil; about 10 percent was found within 10 minutes in the shallow boreholes (not in the piezometers); and the remainder moved rapidly downward in the soil along vertically continuous larger voids, as discussed in Section 1 of this paper. The water level in the holes at 50 cm depth showed a rapid temporary rise. The shallow boreholes form artificial cavities in a three-dimensional system of interconnected, larger natural voids, which conduct the water. These voids may intercept and feed these cavities through their vertical walls. Drainage through their bottom is usually slow, allowing the presence of water during several days, even though the surrounding soil is unsaturated. Water tables should, therefore, not be observed by shallow, unlined boreholes, but by installing tensiometers or piezometers (Bouma et al., 1980).

8

3. THE EFFECT OF TILE DRAINS ON DRAINAGE AND HYDRAULIC CONDUCTIVITY

3.1. Hydraulic conductivity around tile drains

Measurements were made in the heavy clay soil described in Section 2.3. K_{sat}-values below the drain as measured with the drain-cube method had median values of 10 m/d and agreed well with those measured in the undisturbed soil at the same depth, indicating that there were no changes due to either system construction or system operation. The median K_{sat} for the backfill above the drain was 5 m/d. The results pertain to three types of drains, e.g. one clay pipe and two types of plastic pipe. The dye studies showed that water movement occurred almost exclusively along larger planar voids (cracks) and hardly along channels (root holes). Water conducting voids were below 2% by volume (Bouma et al., 1981).

3.2. Hydraulic conductivity of soil above the drains

Sites where tile drainage had been applied for a period of 15 to 20 years were compared with sites without tile drainage (Bouma et al., 1979). Some sites studied by Van Hoorn (1960) were visited again. Also, new drained and undrained sites were selected in 1978 and measurements of K_{sat} were made at all sites with the column method for depths of approximately 30 to 60 cm below surface. Results are summarized in Figure 4.

A statistical comparison of both types of measurements has shown that differences in K_{sat} for undrained conditions was 5 cm·d^{-1} and the value for tile-drained soil was 50 cm·d^{-1}. The difference is thought to be due to increased cracking in the tile-drained soils. Very little additional cracking may have already a strong effect on K_{sat}, as discussed in Section 1 of this paper.

Tile drainage of marine heavy clay soils in the northern part of the

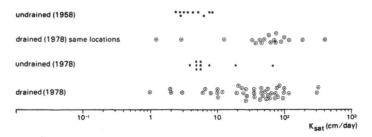

Fig. 4   Measured K_{sat}-values with the column method in tile-drained and undrained heavy clay soils. Values for drained clay soils are significantly higher (after Bouma et al., 1979)

Netherlands also resulted in a significant increase of K_{sat}, but this increase was due to the formation of worm channels (Bouma et al., 1979).

3.3. Drainage of surface soil

Lack of evapotranspiration in the winter period implies that the lowest possible water content in surface soil corresponds with hydraulic equilibrium with the pressure head at every depth is equal to the height above the water table. Equilibrium is not yet reached, however, as long as water flows downwards towards the water table. The natural drainage rate is therefore very important.

Drainage curves for a covered clay soil with a water table at approximately 100 cm below surface (Figure 5) show that drainage rates are very slow. Lowering the water-table levels has no effect on the drainage rate during the first 20 days. This is caused by the particular K-curve of these soils which shows a very strong drop of K upon desaturation (Section 2.2 and Fig. 3). Trafficability and aeration status are only adequate when pressure heads are lower than -90 cm (critical h-value in Figure 5). These data show that natural drainage rates are inadequate to reach sufficiently low water contents in surface soils so that evapotranspiration is needed to significantly lower the water contents (Bouma et al., 1981b).

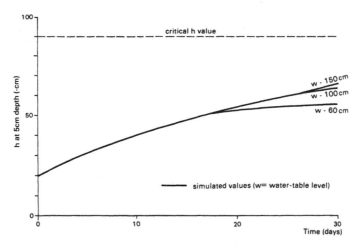

Fig. 5   Slow drainage rate of the surface soil of a heavy clay soil with a constant water table (w) at 60 cm, 100 cm and 150 cm below surface. The initial pressure head was -20 cm. The soil was covered with a plastic sheet to avoid evapotranspiration. The level of the water table has no effect for the first 18 days of drainage. The low drainage rate is due to the low K near saturation (after Bouma, 1981b)

4. REFERENCES

Bouma, J. 1977. Soil survey and the study of water in unsaturated soil. Soil Survey Papers no. 13, Soil Survey Institute, Wageningen. 107 pp.

Bouma, J. 1981a. Soil morphology and preferential flow along macropores. Agric. Water Managem. 3: 235-250.

Bouma, J. 1981b. Soil survey interpretation: estimating use-potentials of a clay soil under various moisture regimes. Geoderma 26(3): 165-177.

Bouma, J. 1982. Measuring the hydraulic conductivity of soil horizons with continuous macropores. Soil Sci. Soc. Am. J. 46: 438-441.

Bouma, J. and Dekker, L.W. 1981. A method for measuring the vertical and horizontal K_{sat} of clay soils with macropores. Soil Sci. Soc. Am. J. 45: 662-663.

Bouma, J., Jongerius A. and Schoonderbeek, D. 1979a. Calculation of saturated hydraulic conductivity of some pedal clay soils using micromorphometric data. Soil Sci. Soc. Am. J. 43: 261-264.

Bouma, J. and Wösten, J.H.M. 1979. Flow patterns during extended saturated flow in two undisturbed swelling clay soils with different macrostructures. Soil Sci. Soc. Am. J. 43(1): 16-22.

Bouma, J., Dekker, L.W. and Haans, J.C.F.M. Haans. 1979b. Drainability of some Dutch clay soils: a case study of soil survey interpretation. Geoderma 22(3): 193-203.

Bouma, J., Dekker, L.W. and Haans, J.C.F.M. 1980. Measurement of depth to water table in a heavy clay soil. Soil Sci. 130(5): 264-270.

Bouma, J., Van Hoorn, J.H. and Stoffelsen, G.H. 1981. The hydraulic conductivity of soil adjacent to tile drains in some heavy clay soils in the Netherlands. J. of Hydrology 50: 371-381.

Bouma, J., Belmans, C., Dekker, L.W. and Jeurissen, W.J.M. 1983. Assessing the suitability of soils with macropores for subsurface liquid waste disposal. J. Environm. Qual. 12: 305-311.

Hoorn, J.W. van. 1960. Groundwater flow in basin clay soil and the determination of some hydrological factors in relation with the drainage system. Versl. Landbouwk. Onderz. 66.10. PUDOC, Wageningen.

Soil Survey Staff. 1975. Soil taxonomy: a basic system of soil classification for making and interpreting soil surveys. Agric. Handbook 436. SCS-USDA, Washington DC.

Factors influencing mole drainage channel stability

G.SPOOR
National College of Agricultural Engineering, Silsoe, Bedford, UK

ABSTRACT

A description is given of the nature of soil disturbance caused by current mole plough designs and the failure mechanisms through which mole channels may subsequently deteriorate. The soil, climatic and implement factors influencing the type of channel failure which actually occurs are discussed.

1. INTRODUCTION

The success of mole drainage installations on fine textured soils of low permeability can be very variable and results are not always consistent on the same soil. Success depends to a large degree on the stability of the mole channel produced and on the associated soil fissures, which allow water to move more freely to the drain (see Leeds-Harrison et al., 1982).

Previous attempts to identify the major factors influencing the success of mole drainage have concentrated mainly on soil and climatic factors. Particular attention has been paid to parameters such as soil texture, stability of soil aggregates to wetting, soil moisture status at the time of moling and timing in terms of season, summarised by Nicholson (1942), Trafford and Massey (1975) and Trafford (1977). Soil classification systems and rules of thumb based on these parameters have been developed to help with the timing and identification of suitable moling situations, but to date they have not been entirely satisfactory. In addition, concentrating only on the soil and climatic factors, which in the main are almost unchangeable, offers little opportunity to improve the moling status of soils which fall into the unsuitable category.

The current design of mole plough used almost universally in the United Kingdom and tried in many other countries, evolved from developments in local areas on a limited range of soils. Whilst this design is likely to be optimum working in situations similar to those where it was developed, it does not necessarily mean it will be optimum for all other conditions. The unsuitability of the plough design could therefore be one reason for grading certain soils into marginal or unsatisfactory categories using existing classifications. Opportunities therefore exist to consider modifications to mole plough design, to improve performance in specific situations where current designs are unsatisfactory.

For mole plough modifications to be successful, a good understanding is first required of the ways in which mole channels deteriorate and fail and how these failure mechanisms are influenced by soil, climatic and implement factors. This paper describes the nature of soil disturbance caused by current mole plough designs, mole channel failure mechanisms identified to date on principally smectic clay soils and the links between the two. The mole plough studies were conducted in both the field and soil bins and the channel failure observations were made using an endoscope on field sites (see Leeds-Harrison et al., 1983).

2. MOLE PLOUGH SOIL DISTURBANCE

Current mole ploughs comprise of three main components (approximate dimensions given), a cylindrical foot (75 mm diameter) with chisel shaped leading edge, a vertical leg (25 mm wide) and a following expander (90-100 mm diameter) (see Figure 1). In operation, the soil responds successively to the action of the foot, leg and finally the expander, the disturbance patterns at each stage being shown in Figure 1. The foot forms a cylindrical

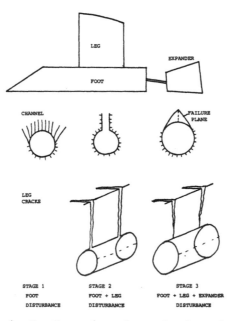

Fig. 1   Stages in the formation of a mole channel and leg cracks

channel by the leading chisel point deforming soil mainly forwards and upwards at an angle of approximately 90° to its face (stage 1). The leg then
splits the channel and creates a series of vertical fissures extending from
the soil surface to the channel, angled up at approximately 45° to the direction of travel, stage 2. Further soil movement is then initiated by the
expander which enlarges the channel by moving soil forwards, upwards and
sideways into the leg slot area along the failure planes shown (stage 3).
Some reduction in the width of the leg fissures also occurs in the vicinity
of the channel.

3. MOLE CHANNEL FAILURE MECHANISMS

Four major types of mole channel deterioration or failure have been
identified to date in the field, each arising under different circumstances,
these are:

- cyclical swell/shrink
- expander
- unconfined swelling
- subsoiler

3.1. Cyclical swell/shrink deterioration (Figure 2)

In channels with a stable roof, channel diameter tends to increase on
drying and decrease on wetting as the soil shrinks and swells. This process
can occur through repeated wetting and drying cycles. Thin sections of the
wall frequently flake off during this process causing debris to build up in
the bottom of the channel. During very dry periods the former leg slot area
may open up and the leg fissures expand allowing soil from the surface layers

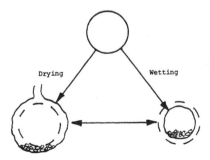

Fig. 2   Cyclical swell/shrink deterioration

15

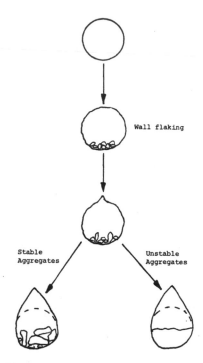

Fig. 3   Expander failure

to fall into the channel. With excessive drying the complete channel may collapse due to the soil peds, loosened by shrinkage, falling in.

3.2. Expander failure (Figure 3)

Where the soil in the roof area is less stable, the whole roof zone tends to drop out leaving a pear-shaped cavity. The roof material slides down slip planes which coincide closely with the failure planes produced by the expander at the time of channel formation (see Figure 1, stage 3). In soils with water stable aggregates, the soil blocks retain their form and water movement can occur readily between them. On the less stable structured soils, the aggregates collapse silting up the bottom of the channel. Further soil peds may fall from the roof area as a result of swelling and shrinkage.

3.3. Unconfined swelling failure (Figure 4)

Many soils will continue to swell into a cavity on wetting, well beyond their swelling limit in an undisturbed cavity free profile. This type of swelling has been termed unconfined swelling (see Spoor et al., 1982). In

16

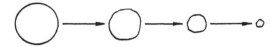

Fig. 4   Unconfined swelling failure

Fig. 5   Subsoiler failure

situations with a stable roof condition, this unconfined swelling around
the mole channel cavity reduces channel diameter, without significantly
changing its shape.

3.4. Subsoiler failure (Figure 5)

In situations where the mole plough has been working close to or above
critical depth (see Spoor and Godwin, 1978) significant soil failure planes
develop from the channel at approximately 45° to the vertical. The implement
causes the soil within these failure planes to move upwards as a unit. With
time this unit settles causing either a flattening of the channel or the
upper channel section moving into the lower section.

4. SOIL/CLIMATE/IMPLEMENT/CHANNEL FAILURE MECHANISM INTERACTIONS

In channels with stable roof conditions, whether deterioration occurs
in the form of cyclical swell/shrink or unconfined swelling failure is dependent upon the weather patterns and the swelling and shrinkage characteristics of the soil. These factors will also influence the rate of deterioration.

The achievement of a stable channel roof when an expander is used, appears to be very dependent upon the nature of the final soil condition produced by the expander. Under conditions where the disturbed soil within the
failure planes (Figure 1, stage 3) becomes strongly bonded with the undisturbed soil beyond, a stable roof condition will develop. Where this bond
is weak, the collapse risk will be high. The strength of bond at this interface will be very dependent upon the soil moisture status and the loading
at the time of channel formation.

Soil packing state, moisture status and implement geometry influence the position of critical depth and hence control whether a subsoiler type failure is likely or not.

The timing of the first wetting of the channel relative to its formation has a significant effect on the rate of deterioration with all types of failure mechanism. Soil disturbance at the time of channel formation disrupts many bonds which help restrict soil swelling on wetting and time is required for these bonds to reform. Rapid wetting after moling is likely, therefore, to cause greater swelling and more rapid channel deterioration.

ACKNOWLEDGEMENTS

The author wishes to express his thanks to the Agricultural and Food Research Council and the Ministry of Agriculture, Fisheries and Food for their financial support.

REFERENCES

Leeds-Harrison, P., Spoor, G. and Godwin, R.J. 1982. Water flow to mole drains. J. Agric. Engng. Res. 27: 81-89.
Leeds-Harrison, P.B., Fry, R.K., Cronin, C.J. and Gregory, J.E. 1983. A technique for the non-destructive monitoring of subsurface drains. J. Agric. Engng. Res. 28: 479-484.
Nicholson, H.H. 1942. The principles of field drainage. Cambridge University Press. 165 pp.
Spoor, G. and Godwin, R.J. 1978. An experimental investigation into the deep loosening of soil by rigid tines. J. Agric. Engng. Res. 23: 243-258.
Spoor, G., Leeds-Harrison, P.B. and Godwin, R.J. 1982. Some fundamental aspects of the formation, stability and failure of mole drainage channels. J. Soil Sc. 33: 411-425.
Trafford, B.D. and Massey, W. 1975. A design philosophy for heavy soils. Tech. Bulletin 75/5, Field Drainage Experimental Unit, Cambridge.
Trafford, B.D. 1977. Recent progress in field drainage. Part I. J. Royal Agric. Soc. of England 138: 27-42.

Effective disruption is a major factor in the drainage of impermeable soils

L.F.GALVIN
Kinsealy Research Centre, Dublin, Ireland

ABSTRACT

Water table measurements at two sites (Kanturk and Ballyroan) show a deterioration in water table control on mole drained plots when compared to gravel moled plots. The deterioration which is much more pronounced at Kanturk is attributed to the combined effects of a breakdown of drainage channels and of the vertical angled crack system. For the effective drainage of impermeable soils the provision of stable long-lasting channels is a major priority. Field investigations and theoretical considerations highlight the importance of a well-developed crack system and indicate the need for investigating the re-design of mole drainage equipment.

1. INTRODUCTION

In Ireland, the average annual rainfall varies from about 800 mm along the East coast to more than 1500 mm on the West coast. The annual potential evapotranspiration from grassland is of the order of 400 mm. The rainfall surplus therefore ranges from 400 mm to more than 1100 mm annually. Another aspect of Irish rainfall is that on average 40 to 45% of the annual total falls during the April to September period. This rather evenly distributed rainfall throughout the growing season coupled with the shallow topsoil depth of most of the impermeable soils results in most of these soils being used for grassland production.

In the undrained condition, major trafficability problems arise for both animals and machinery. Early nitrogen application is very often difficult and sometimes impossible and during wet summers, the land can become impassable for machinery with the result that forage is either lost or else harvested with difficulty and corresponding cost increases. In similar circumstances the grazing animals poach the soil surface badly and much of the grass is wasted. In both cases grass production for the following year is adversely affected. The problems are accentuated when greater than normal rainfall occurs over a number of consecutive growing seasons. This occurred in parts of the West of Ireland in the 1979-1981 period, and resulted in many fields becoming totally impassable for animals and machinery, leading to an enforced reduction in cow numbers and a corresponding income drop on many farms. In such circumstances the benefits of drainage must be considered in the context of both production and utilization, rather than in the context of production only.

1.1. Impermeable soils

In a survey on land drainge problems and drainage installations in Ireland (Galvin, 1969), it was found that 31% of all drainage problems in the country were classified as impermeable soils. Of these, the most important group consists of the glacial tills. These are usually overconsolidated to varying degrees and are also very impermeable with conductivities generally less than 1 cm/day.

1.2. Drainage systems

Traditional pipe drainage systems even if installed at uneconomically close spacings do not provide adequate drainage in these heavy impermeable soils. It is therefore necessary to resort to drainage systems that incorporate both soil disruption and the formation of drainage channels at relatively close spacings (1.3 to 2 m). The cracking and loosening achieved during soil disruption facilitates the transfer of water from the surface to the drainage channels, through which it is then transported to the pipe collector drains. The long-term effectiveness of such drainage systems obviously depends on the effectiveness and permanency of the crack structure and channel system. Mole drainage has been used in many parts of the world, over a number of years, as a method of providing a relatively inexpensive drainage system for impermeable soils. If installed properly it can be a very effective and efficient system of drainage where the soils are sufficiently stable to maintain an open crack structure and a mole channel which continues to discharge water over a period of 5 to 10 years. However, many soils are not stable.

In Ireland mole channels have collapsed completely within a period of 1 to 2 years after installation. Arising from this the gravel-mole system was developed by Mulqueen and Prunty and has been installed on a large number of farms in Northern Ireland and in the Republic of Ireland. The system is very effective in providing stable long lasting channels.

Pilot drainage trials incorporating mole drains ripping and gravel moles were installed at a number of sites in the West of Ireland in the period 1976-1978. The results of these trials were encouraging and on some sites, spectacular evidence was provided of widespread damage to the soil surface, arising from the breakdown of particular drainage treatments (Galvin, 1982). In 1981-1982 under a project, part-funded by the EEC, a number of instru-

20

mented experimental trials were installed (Galvin, 1983). Data from two of these trials (Kanturk and Ballyroan) are outlined and discussed.

2. EXPERIMENTAL SITES

Kanturk is situated in County Cork in the South West of Ireland whereas Ballyroan is located in County Laois in the South East midlands. Details of the rainfall pattern at each site and of major soil physical properties are given in Tables 1 and 2.

TABLE 1   Annual average rainfall (1951-1980) for Kanturk and Ballyroan (mm)

Month	Kanturk	Ballyroan
January	118	94
February	83	69
March	79	64
April	60	61
May	68	74
June	53	58
July	62	68
August	72	74
September	92	87
October	96	88
November	107	86
December	123	99
Total (annual)	1013	922
Total (April-September)	407	422

TABLE 2   Particle size distribution, Atterberg limits and bulk densities (Mg/m^3) of subsoils

	Kanturk	Ballyroan
Percentage passing 2 mm	100	100
0.6	97	96
0.2	92	90
0.06	77	77
0.02	65	65
0.006	50	45
0.002	34	31
Percentage >2 mm	9	5
Clay/silt ratio	0.52	0.48
Liquid limit	37	44
Plastic limit	21	25
Plasticity index	16	19
Bulk density	1.50-1.62	1.61-1.78

21

The rainfall data (Table 1) show that whereas the annual average rainfall at Kanturk exceeds that at Ballyroan by 90 mm, the rainfall during the growing season (April-September) at both locations is very similar. Both sites are however representative of the drier areas on which research on the drainage of impermeable soils is being carried out.

Table 2 shows that the physical properties of the soils are not very different. The particle size distribution data are practically identical over the 2-0.2 mm range and only vary to a maximum of 5% thereafter. The Atterberg limits are quite similar and clay/silt ratios are typical of many impermeable Irish soils.

On both sites similar disruption treatments are installed. These include:

- mole drains spaced at 1.3 m;
- gravel moles spaced at 1.3 m;
- gravel moles + ripping. The gravel moles are spaced at 2.6 m and ripping is then carried out between adjacent gravel moles;
- control.

The plots at Kanturk are 900 m^2 and those at Ballyroan 800 m^2. At each site the water flow from the individual plots is led through separate collector drains to continuous flow recorders. Each site is also equipped with a recording rain gauge and with maximum reading piezometers and/or water level recorders as detailed below.

In considering the water level data outlined in the paper, it should be remembered that the soil is not homogeneously disrupted. The maximum cracking and fissuring occurs in the immediate vicinity of the disruption channel and the effectiveness of disruption decreases as the distance from the disruption channel increases (Leeds-Harrison et al., 1982). All the water level recorders and piezometers are installed midway between adjacent disruption channels. The water level fluctuations are therefore measured at points where the cracking and consequently the drainage effect is minimised.

2.1. Kanturk

The pipe collector drains were installed at Kanturk in June 1981 and the disruption treatments in August 1981. The flow recorders and maximum reading piezometers were installed in March 1982. Water level recorders were installed on each plot in 1983.

2.2. Ballyroan

The collector drains were installed at Ballyroan in July 1982 and the disruption treatments in August 1982. Flow recorders and maximum reading piezometers were installed on all plots in February 1983 and water level recorders added in November 1983.

3. SITE INVESTIGATIONS AND RESULTS

The discharge and water table level charts are collected weekly. The maximum and current water levels in the maximum reading piezometers and the ground conditions are observed at the same time. The condition of the moles, gravel moles and rip channels is investigated by excavation and channel casting (where possible) at regular intervals.

3.1. Site investigations at Kanturk

Detailed investigations were carried out at the Kanturk site in June 1982, August 1983 and May 1985, during which randomly selected moles, gravel moles and rip channels were excavated. The condition of the channels and the crack structure in their vicinity were examined. However due to the rate at which the mole and rip channels had slurried up on this site, it was not possible to take polyurethane casts on any of the three occasions.

All the gravel moles examined were in good condition. The leg slot was invariably filled with topsoil which had migrated down to the top of the gravel and extended from there to the base of the topsoil horizon. This band of topsoil (up to 50 mm wide) was generally loose and permeable. The gravel in the mole had not been contaminated by the topsoil in the leg slot or by the surrounding subsoil. The vertical angled cracks which were observed in 1983 were approximately 0.25-0.30 m long, 6-9 mm wide and spaced at 0.20-0.25 m. These cracks were also found in 1985 but were then less distinguishable. This may be due to the fact that wereas the 1983 excavation was carried out at the end of a long dry period in August, the 1985 excavations were undertaken in early May when the soil moisture deficit was probably close to zero. However, the possibility that the width of the cracks is decreasing with time cannot be ignored.

The rip channels examined in 1983 were filled with topsoil. Topsoil was also found to varying heights in the leg slot. In 1985 the channels examined were filled with topsoil as were the leg slots. The band of topsoil in the leg slot was up to 60 mm wide. It was relatively uncompacted and capable of transporting water. The ripping had been carried out by the same machine as that used to install the gravel moles. In those circumstances the vertical angled cracks found radiating from the rip channels were similar to those found on the gravel mole plot.

The moles excavated in 1982 were filled with a loose slurried material. They appeared to be quite capable of transporting water to the collector drains. The moles excavated in 1983 were more solidly clogged. Occasional gaps were found between this infill material and the original wall of the mole channel. Three moles were examined in 1985. Two were completely clogged with material (mainly subsoil and some stones, with occasional traces of topsoil). The third mole was also clogged but had a slight (10 mm high) gap between the top of the infill material and the subsoil above it. This gap did not appear to be continuous but was observed at a number of points along the mole. The dimensions of the plug of infill material varied from 80 x 80 mm to 150 x 80 mm. This indicates that although the width of the original channel has not altered, substantially the vertical dimension has changed. On examination it would appear that the soft material extends below the original channel (due to water softening) and upwards due to the collapse of the roof. The problem in this regard at Kanturk are accentuated by the stony nature of the subsoil. This is borne out by the number of stones up to 30 mm diameter found in the infill material.

The leg slot crack was sometimes difficult to distinguish within the soil mass but, on excavation, the soil mass usually sheared open along this vertical plane. In one mole, however, the leg crack was readily observed and in another mole a very narrow band of topsoil indicated its position. The vertical angled cracks were also difficult to find on many moles but were very evident in the mole on which the narrow band of topsoil had been observed.

During the 1985 excavations, a very slight seepage of clear water was observed in most moles. However, the general impression created was that the moles are almost totally clogged up.

3.2. Site investigations at Ballyroan

Excavations were carried out at Ballyroan on 3 May 1985, during which the crack structure and the condition of the various disruption channels were checked.

The gravel moles are in good condition. There has been no contamination of the gravel in the mole by the surrounding soil. Topsoil has migrated into the wide leg crack to the top of the gravel mole and provides a direct connection from topsoil horizon to the gravel. The cracking and shattering of the subsoil in the vicinity of the gravel mole is also very good.

The cracking around the rip channel is well-developed. However, the channels were almost completely blocked up by a combination of subsoil swelling and the ingress of topsoil through the leg crack. In one of the rip channels examined, there was a considerable amount of topsoil in the leg crack and rip channel. This was generally approximately 20 mm wide but at one point had expanded to a diameter of 45 mm to fill a large hole about 12 cm above the channel base. The topsoil was relatively uncompacted and seemed quite capable of transporting water to the collector drains.

The mole drains were also in good condition although two blockages were found in those examined. One blockage had been caused by the disturbance of some large stones which resulted in a large quantity of soil being deposited in the channel, blocking it completely. A minor roof collapse appeared to be the main cause of the second (partial) blockage of the channel. The general average channel section observed was approximately 65 mm x 45 mm, ranging from a minimum of 25 x 25 mm to a maximum of 70 x 45 mm. The soft material around the channel varied in overall dimensions from 70 x 70 mm to 130 x 80 mm.

At Ballyroan the vertical angled cracks radiating from the moles were readily observed. They were well-developed with roots extending right down the cracks to the full depth, and were spaced approximately 0.17 m apart. Polyurethane casts were taken on three moles. The two blockages already referred to, were discovered during casting.

3.3. Water table fluctuations

The disruption treatments were installed at Kanturk in August 1981 and at Ballyroan in August 1982. The water table fluctuations over the September–May period for 1982/83 and 1983/84 (Kanturk) and for 1983/84 and 1984/ 85 (Ballyroan) are analysed.

The weekly water table levels at Kanturk for 1982/83 (Figure 1) and for 1983/84 (Figure 2) show quite clearly the deterioration in water table control that has occurred on the moled and gravel moled + rip plots relative to the gravel moled plot over that period. This deterioration is borne out by the SEW data outlined in Table 3. The SEW concept which is a measure of the water table height and duration above a given level was defined by Sieben (1964) and discussed by Wesseling (1974).

25

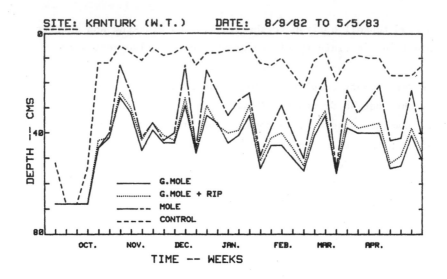

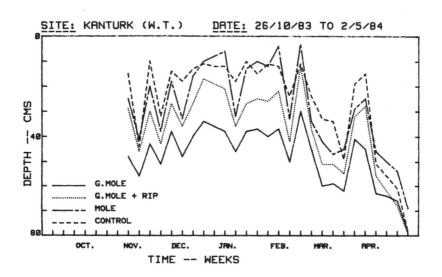

TABLE 3 SEW (30) figures for the September-May periods at Kanturk
(1982/83 and 1983/84) and at Ballyroan (1983/84 and 1984/85)

	Kanturk		Ballyroan	
	1982/83	1983/84	1983/84	1984/85
Gravel moles	13	0	111	0
Gravel moles + rip	35	441	420	139
Moles	520	1.389	84	63
Control	4.042	1.751	1.961	1.492

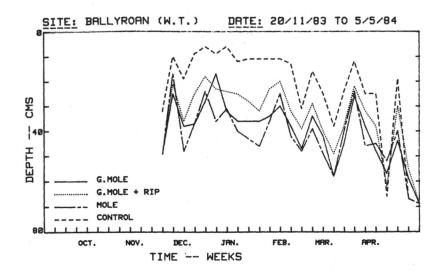

SITE: BALLYROAN (W.T.)   DATE: 20/11/83 TO 5/5/84

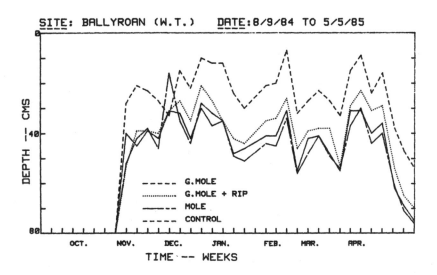

SITE: BALLYROAN (W.T.)   DATE:8/9/84 TO 5/5/85

The Ballyroan water table data are shown in Figure 3 (1983/84) and
Figure 4 (1984/85). The SEW figures are detailed in Table 3. These data show
that the water table control in the gravel moled + rip plot is slightly
less effective than in the gravel moled plots. There seems to be some indication also that the mole drained plot is beginning to show slight signs of
deterioration relative to the gravel moled plot. However, this is not
clearcut at this stage, and the mole drainage is still very good.

27

4. DISCUSSION

At Kanturk the deterioration in water table control of the mole drains relative to the gravel moles could be attributed to the very obvious breakdown of the mole channels. Whether there is a corresponding reduction in the effectiveness of the crack system is not clear but it is reasonable to assume that the crack structure has also deteriorated to some extent. The slight fall-off in mole drainage efficiency at Ballyroan might also be attributed to crack deterioration since the mole channel system there is largely intact. In fact, the moles at Ballyroan show far less deterioration than those at Kanturk over the 3-year period.

The deterioration in water table control on the gravel moled + rip plots relative to the gravel moled plots at Kanturk appears to be mainly due to the infilling of the rip channels and some consequential crack deterioration in their vicinity.

It therefore appears that the deterioration in drainage effectiveness can be attributed to a combination of channel breakdown and crack closure. On that basis the priorities in disruption drainage are: 1) the provision of stable long-lasting channels for water transport and 2) the development of an adequate crack structure in the soil.

The transfer of water from the soil surface to the drainage channels depends to a large extent on the effectiveness of: a) the leg slot crack and b) the vertical angled cracks both of which are governed to some degree by the width of the leg of the disruption equipment. The wider leg (approx. 90 mm) of the machine used to install the gravel moles and ripping produces a much wider leg crack than that produced by the 25 mm wide leg of the mole plough. Site investigations have shown that the wider leg slot made by the gravel mole machine is very often filled with relatively loose topsoil whereas only traces of topsoil are found along some of the leg slots made by the mole plough. The wider-legged equipment therefore tends to produce a much more permeable and longer lasting leg slot. This is very useful when the machine is used to install gravel moles. However, where the machine is used as a ripper and the channel is not filled with gravel, the wide leg slot allows the channel and slot to fill with topsoil, thus reducing the flow capacity of the channel drastically.

The wider leg also has an advantage in the production of the vertical angled dracks, which tend to be wider and somewhat longer where the gravel mole machine is used. This is confirmed by soil in studies carried out at Silsoe College (Spoor, 1983) using vertical tines of differing widths in an artificial clay soil. This study showed that the size of cracks associated with the leg slot increased as the width of the leg increased. When the roughness of the leg was increased by glueing coarse sand to both faces, an increase in crack volume was measured. It was found that a 6 mm roughened tine could give a volume of cracks equivalent to a 25 mm smooth tine. However, the 6 mm roughened tine produced a large number of small cracks whereas the 25 mm smooth tine produced a smaller number of wide cracks.

Youngs (1984) has shown that the length of the angled vertical cracks and to a lesser extent their spacing are major factors in the effectiveness of mole drainage. It would appear therefore both from theoretical considerations and experimental investigations that the length, spacing and rate of deterioration of these vertical cracks is of vital importance in the drainage of impermeable soils by disruption methods. As the rate of crack deterioration is probably related to the initial crack width, this is also a factor. The re-design of mole drainage equipment to produce a stable long-lasting channel and to maximise on the production of the most effective combination of length, spacing and width of vertical angled cracks under a variety of soil conditions would therefore seem to be a priority.

ACKNOWLEDGEMENT

The author acknowledges the assistance of M. O'Herlihy, P. Healy, P. McCormack and L. Foley. Site readings at Kanturk were carried out by E. Walsh, J. O'Donovan and T. Weldon, ACOT and at Ballyroan by P. McEvoy. The project was part-funded by the EEC under contract no. 0582. The assistance of the commission is gratefully acknowledged.

REFERENCES

Galvin, L.F. 1969. Land drainage survey - 11. Ir. J. Agric. Res. 8: 1-18.
Galvin, L.F. 1982. Gravel mole drains are the most suitable system for many impermeable soils. Fm. Fd. Res. 13: 103-105.
Galvin, L.F. 1983. The drainage of impermeable soils in high rainfall areas. Ir. J. Agric. Res. 22: 161-187.
Leeds-Harrison, P., Spoor, G. and Godwin, R.J. 1982. Water flow to mole drains. J. Agric. Engng. Res. 27: 81-91.
Sieben, W.H. 1964. Het verband tussen ontwatering en opbrengst bij de jonge zavelgronden in de Noordoostpolder. Van Zee tot Land 40. Tjeenk Willink V, Zwolle.
Spoor, G. 1983. Personal communication.
Wesseling, J. 1974. Crop growth and wet soils. In 'Drainage for agriculture' (ed. J. van Schilfgaarde). Agronomy 17: 7-37, Amer. Soc. Agron., Madison, Wisconsin.
Youngs, E.G. 1984. An analysis of the effect of the vertical fissuring in mole-drained soils on drain performances. Agric. Water Management 9: 301-311.

Discussion Session 1

Question from Mr. J. Mulqueen to Mr. J. Bouma

    Has Mr. Bouma used the pit bailing method of Bouwer and Rice to measure
K_{sat}? We have had encouraging results with this method. Have Mr.
Bouma's results been good?

Answer of Mr. Bouma

    Yes, this method yields good results. However, it uses a relatively
large volume of soil and is quite laborious. We try to reduce the
amount of work as much as possible. In our clay soils we could reduce
sample size to about 15 litres with samples still being representative.
Smaller samples were not.

Question from Mr. P.J.T. van Bakel to Mr. J. Bouma

    Is it not better to use pressure transducers instead of piezometers to
measure groundwater depths (phreatic levels), because no water movement is involved?

Answer of Mr. Bouma

    I agree. We did so in the study reported in Soil Science. Tensiometers
will indicate water pressure inside the pods. Water movement through
macropores follows independent patterns, that are very hard to trace.

Question from Mr. J. Feyen to Mr. J. Bouma

    From your paper we learned how 'Dutch' heavy clay soils are character-
ized. However, how do we integrate these properties in existing water
balance models (like e.g. the SWATRE model) so that we can cope with
bypass flow phenomena from the top layers to the underground, in some
cases even to the groundwater? We faced this problem by analyzing field
data which were monitored in the Nile delta, near Alexandria, where the
site is characterized by a swelling-cracking heavy clay and a shallow
water table. Is there anything what we can do?

Answer of Mr. Bouma

    We have attempted to approach this problem by considering the soil as
a bi-porous system. Flow through macropores occurs when the small
pores in the soil mass between the macropores cannot handle the flow.
The amount of bypass flow (= flow of free water through an unsaturated
soil mass) can be measured with a field method. It values as a function of the moisture content, microrelief, rain intensity and duration.

Using such relationships, bypass flow can be quantified and the moisture regime of bi-porous soils can be simulated successfully. Reference is made to the proceedings of the International conference on water and soluble movement in heavy clay soils (eds. J. Bouma and P.A.C. Raats), ILRI Publication, Wageningen, the Netherlands. Mr. Armstrong's paper, to be presented later, gives a comparable approach.

Question from Mr. Ph. Lagacherie to Mr. G. Spoor

Our field observations on experimental sites show a lot of examples of what you call expander failure. We have observed in the channel aggregates coming from the wall of the channel. These aggregates seemed to us more stable than those from the undisturbed soil and we attributed this apparent stability to the expander work. On the other hand, we saw that there were a lot of voids between these aggregates so that the water can flow through the channel as it does in a gravel mole channel. Can we expect that an increase in expander diameter leads to a better water aggregate stability, giving us a method for increasing the life of the channels?

Answer of Mr. Spoor

These flat aggregates come from the sides of the channel rather than from the roof. They are formed by the reworking of the soil at the mole foot/soil interface. The amount of reworking is dependent upon the sliding resistance at the interface. This resistance changes with moisture content as well as with the roughness of the mole foot, but at the sides of the channel, it is not significantly affected by foot or expander diameter. Expander diameter is therefore not likely to have a large influence on these flat aggregates. The main effect of expander diameter is in the roof of the channel where it can influence soil packing and hence roof stability.

Question from Mr. H. Wösten to Mr. G. Spoor

What determines at what depth the mole drain should be created in the soil profile?

Answer of Mr. Spoor

The depth chosen depends upon:

- weight of traffic carried on surface (deeper in arable than grassland);
- depth of most suitable moling soil in profile;
- power unit available to pull mole plough under the prevailing moisture conditions;

- likely water deficit developed during drier seasons. The greater the potential deficits, the deeper the moles so that swelling and shrinkage in the channel itself can be minimized.

Common moling depths in UK range between 0.4 and 0.6 m. Interest is increasing in shallower depths where a longer life is not so essential.

Question from Mr. J. Bouma to Mr. G. Spoor
- Taking your examples of mole-channel stability, has it been tried to form arch-shaped channels that would be more stable structurally than round ones?
- Is mole-channel effectivity closely correlated with soil types or is variability of results within each soil type so high so that it obscures differences between soil types?

Answer of Mr. Spoor

No systematic comparisons have been made between different shapes of channel. It would be worth looking at an arch-shaped channel in situations where roof collapse through expander type of failure is likely to occur. It would be unlikely to increase stability in situations where unconfined swelling or cyclic swell/shrink are the major failure mechanisms.

It is possible to identify good moling soils and bad moling soils on a basis of soil type and soil series. The major problem is with the intermediate marginal soils where variability within the soil type and between different seasons can exceed that between soil types. It is hoped that implement modifications will eliminate some of these variations, allowing a better ranking, in terms of moling potential, to be made for these marginal soils.

Question from Mr. B. Lesaffre to Mr. G. Spoor

You described four failure mechanisms. In each case, the rate of channel collapse must depend on soil type, moisture content, type of implement and so on, as you told us. My questions are:

- is it possible to define the very moment, at which the channels have become no longer efficient?
- have you got an idea of channel lifetime?

Answer of Mr. Spoor

The most useful way of assessing the efficiency of the channels is to examine the shape of the discharge hydrograph. Newly installed efficient mole systems have a very peaky hydrograph following a storm. The peak decreases, for the same storm, as the moles deteriorate.

Channel life varies between soils and on the same field. On the same field installation, moisture content and the following weather pattern influence life. Channel wetting soon after installation causes more rapid collapse. If a channel can mature for 2 to 3 weeks before being wetted, it will have a longer life. The aim in future is to adjust the implement to give the most stable channel possible under the prevailing conditions. In UK, farmers on the good moling soils expect to re-mole on average once every 5 years. On the poorer moling soils, re-moling is carried out every 2 to 3 years.

Question from Mr. P. Bazzoffi to Mr. L. Galvin

Did you find significant differences in grass production using normal mole and gravel mole drainage? What is the best size for gravel pieces?

Answer of Mr. Galvin

There are no major differences in production between gravel moled and successfully mole (natural) drained land. However, when mole drains begin to fail there are major problems of trafficability for animals' grazing and silage harvesting. The big problem is therefore one of utilization of the grass. Where mole drains fail within a short period, there is a major risk of crop loss and of large-scale damage to the soil surface.

Gravel size for gravel moles is 12-20 mm.

Question from Mr. B. Lesaffre to Mr. L. Galvin

You showed us how efficient a gravel mole system is compared to plain moling. Have you got an idea of the cost effectiveness of both gravel mole and plain mole drainage networks, including, in the latter case, the channel renewal every third or fifth year?

Answer of Mr. Galvin

The costs of gravel moles and ordinary moles in Ireland are in Ir. £ (approximately equal to US $), per hectare:

moles:		gravel moles	
collector drains	Ir. £600	collector drains	Ir. £ 400
moling	Ir. F100	gravel moling	Ir. £ 700
	Ir. £700		Ir. £1100

The cost of re-moling by a contractor is approximately Ir. £200-250 per hectare.

For tillage soils where there is a possibility of damaging the crack structure during cultivation or harvesting, repeated moling may be most cost effective. The same situation exists for grassland where the fields are grazed. However, on stony soils it is often necessary to re-level and re-seed land for silage harvesting. Gravel moles are therefore more cost effective for silage land.

Question from Mr. J. Wesseling to Mr. G. Spoor and Mr. L. Galvin

Can moling be considered to be a normal farm practice or is it to be performed by contractors that have the proper machinery available? This generally will make a difference when considering costs.

Answer of Mr. Spoor

Traditionally in UK, moling has been regarded as a contractors operation. With the increase in power available on arable farms, however, it is now becoming more of a farm operation. Mole drainage has much more potential if it can become a farm operation. Silsoe College is investigating ways of making stable moles at shallower depths to reduce the draught and traction requirements. This will allow smaller powered tractors, now available on farms to be used.

Answer of Mr. Galvin

In Ireland mole drains installed at a depth of 0.5 m is generally a contractor operation. In recent years, however, good results have been obtained from shallow moles (approximately 30 cm deep) as an ancillary treatment over existing moles or gravel moles installed at 0.5 m depth. These shallow moles can be installed by the farmer using a mounted mole plough and a low-powered tractor.

Question to Mr. J. Bouma and Mr. J. Wesseling

Why has mole drainage not been applied more extensively in the Netherlands? Is it due to lack of sloping land or traditional emphasis on tile drainage?

Answer of Mr. Bouma and Mr. Wesseling

Maybe lack of slope may have been a factor in not applying moling. Soils like the river basin clays, suitable for moling and used as grassland were drained by means of furrows. In most areas subsoils are more sandy and not suitable for moling. In addition there have always been wet conditions due to elevation so subsoils were less suitable for moling.

Question from Mr. A. Armstrong to Mr. Ph. Lagacherie

You mentioned use of soil as backfill in place of expensive permeable backfill. Our experience is that natural soil as backfill is initially sufficiently permeable in nearly all cases, but this permeability decreases with time, so that after 10 years the soil is fully consolidated and the drain does not react to moling or subsoiling above it. Because of this, we now recommend the use of permeable backfill in all circumstances where the drainage includes mole drainage. Have you any experience with this decline of permeability of natural soil in the trench?

Answer of Mr. Lagacherie

As I said drainage of heavy soils is not yet very old in France. We just have linked field observations with farmers appreciation. This shows in very particular clay soils (alluvial clay soil, stratified marsh clay soil) a good permeability of five years old trenches. So we think that drainage with natural backfill and a close drain spacing (10-15 m) without moling or subsoiling is a suitable drainage technique for these clay soils. But when we have clay soils that need moling we put also gravel in the trench.

I have observed in England that packing problems of natural backfill just below the ploughing layer may occur, while natural backfill located near the drain pipe is still porous. If this phenomenon cannot be avoided, it is obvious that natural backfill is not suitable in all clay soils.

Session 2
Effects of drainage and/or irrigation on agriculture

Chairman: H.C.ASLYNG
Hydrotechnical Laboratory, RVAU,
Copenhagen, Denmark

Twenty years of drainage and irrigation research

S.DAUTREBANDE
Faculté des Sciences Agronomiques de l'Etat, Gembloux, Belgium

ABSTRACT

The Hydraulic Department (Génie Rural II) of the State Faculty of Agronomy at Gembloux has been conducting research on drainage and irrigation problems for twenty years. An overview of the results of this research is given in order to identify certain areas for further investigation. The most important results are compiled, and the synthesis is presented largely in tables and graphs.

1. AGRICULTURAL DRAINAGE STUDIES

1.1. Studies

The studies were begun in the nineteen sixties and have evolved along the following lines:

Measurements

The registration of rainfall and the height of the groundwater level under grasslands with alluvial clay and sandy soils. The drained and non-drained conditions were compared as well as draining with subsurface drains at different spacings. Two examples are presented in Figures 1 and 2 (drains at a depth of 0.60 m). In general, one may conclude that there is a significant effect of drainage in wet periods and no drainage in dry periods (lowering of the groundwater level below the drains) (Report IRSIA 1965-1967).

From measurement of fodder yields in relation to the drainage situation it was noted that drainage is favourable in heavy to medium-heavy soils and unfavourable when too excessive in light low retention soils (Figures 3, 4 and 5) (Report IRSIA 1967-1969).

Interpretations

Studies of the duration of saturation of soils at different depths (Figure 6, Report IRSIA 1963-1965). The interannual variability was established which shows the necessity of statistical pluviometric studies. Analysis of the duration of saturation of the soil as a function of the spacing between drains leads to the conclusion that this kind of study should be developed

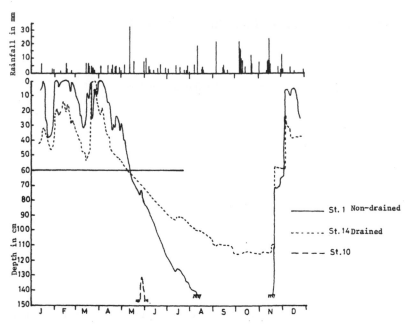

Fig. 1   Fluctuations of groundwater levels, Famenne 1964

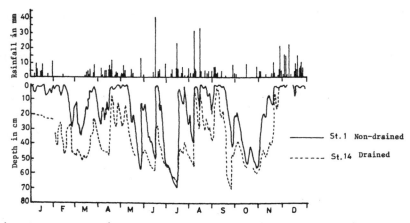

Fig. 2   Fluctuations of groundwater levels, Famenne 1965

further. Study of drainage porosity (L. Sine and S. Dautrebance-Gaspar, 1968) and of the variability of hydraulic conductivity (Calembert, 1962) were carried out as an example of the results based on the analysis of limni-graphs of the groundwater level is given in Table 1.

On the basis of finite differences and finite elements numerical models formulas were developed for the spacing between tile lines of buried drains

40

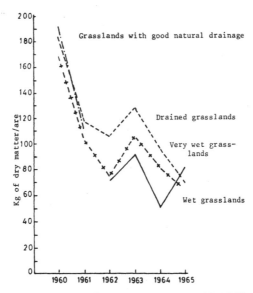

Fig. 3   Average grassland yields, Hesbaye 1960-1965

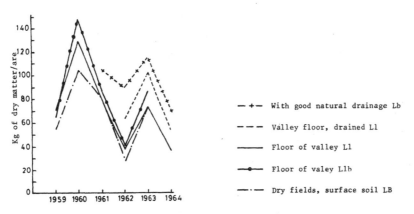

Fig. 4   Average grassland yields, Famenne

in homogeneous soils (Sine and Vermeiren, 1965) and for stratified soils (Sine and Dautrebance-Gaspar, 1978). An example is given in the form of the nomograph of Figure 7.

A comparison between steady and non-steady state drainage formulas was carried out (see Figure 8) (Sine and Dautrebance-Gaspar, 1968).

1.2. Areas for further investigation

The interannual variability of the results demonstrates the importance of mathematical models to link drainage performance with rainfall events in

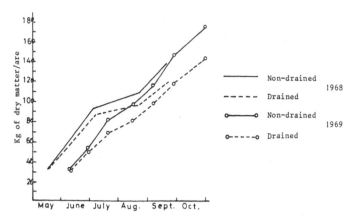

Fig. 5   Average grassland yields, Campine

order to be able to derive a basis for statistical and economic considerations. It also follows from the great diversity in drainage formulas and
from application of them to field data (effect of drain spacings) that it
would be useful to establish systematic field studies. Finally, study of
the quality of the drainage water and the drainage effects on surface flow
represent equally attractive research objectives.

2. STUDY OF IRRIGATION

2.1. Studies

The importance of supplementary irrigation for certain soils and certain crops in Belgium has been demonstrated and has led to the development
of a pilot project (Ben Harrath et al., 1981).

The trials in open fields involved the comparison of irrigated and non-
irrigated conditions with measurements of water and tensiometric profiles.
The results can be briefly summarized as follows:

- There is no difference in water consumption of grasslands when levels of
  nitrogen fertilizer are changed (Figure 9) (Ledieu et al., 1985).
- The uptake of nitrogen by crops varies with the supply of nitrogen fertilizers (Table 2) (Report IRSIA 1977-1979).
- The effect of irrigation is related to the effect of the supply of fertilizer (Figure 10) (Sine et al., 1979).
- In periods of prolonged dryness, soil water profiles develop that are
  characteristic of the individual crops (Figure 11) (Ben Harrath et al.,
  1977).

42

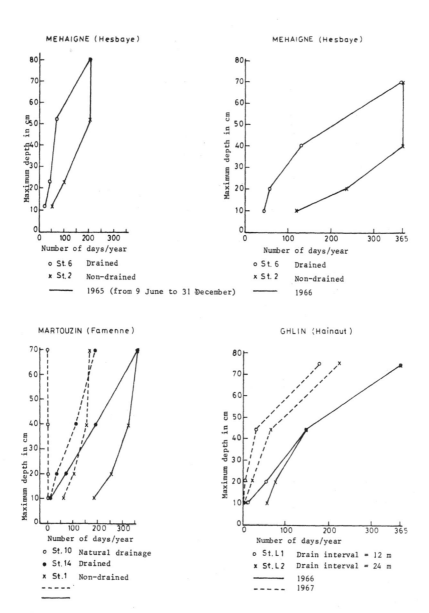

Fig. 6   Periods of saturation of the soil by groundwater

TABLE 1   Values of P

Station	Lower limit of h in cm	h_1 in cm	a	Calculation of P
5	18	47.5	0.0336	0.028
6	18	44.7	0.0418	0.029
14	–	51.0	0.0088	0.027
27	15	41.5	0.0097	0.024

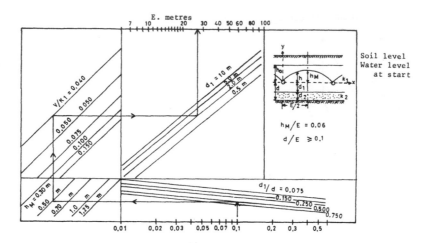

Fig. 7   Nomograph for the dimensioning of drainage in stratified soils
with $K_2 > K_1$, $d/E > 0.1$, ahd $h_M/E \leqslant 0.06$

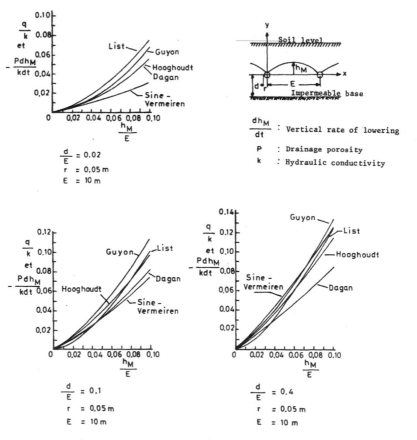

Fig. 8   Comparison between the variable and the permanent systems

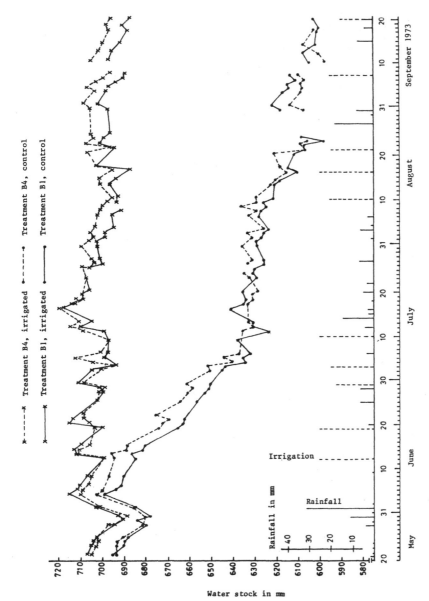

Fig. 9   Temporary grassland: ray grass. Evolution of water stocks observed in situ
(average of 3 profiles)

TABLE 2   Nitrogen balance: supply and removal of kg of nitrogen per hectare. Irrigation trial, Gembloux, 1976

Treatment	200A	466A	466C	600B	
Control	66	312	292	522	supply (fertilizer)
	93	277	210	499	removal
	-27	+35	+82	+23	balance
Irrigation 1	132	544	521	756	supply
(deficit = 20 mm)	174	547	524	899	removal
	-42	-3	-3	-143	balance
Irrigation 2	132	544	521	756	supply
(deficit = 60 mm)	184	551	536	(824*)	removal
	-52	-7	-15	(-68)	balance

*'burning' phenomena

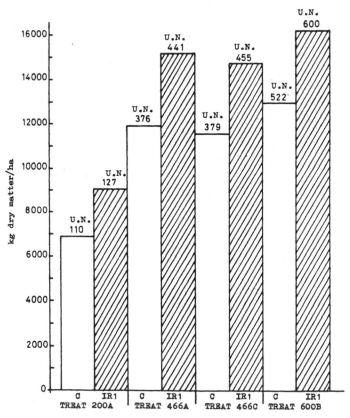

Fig. 10   Irrigation trial, grasslands, Gembloux
Histogram of average annual production from 1972 to 1977 (kg dry matter/ha)
U.N. = units of nitrogen fertilizer, annual average
C = control
IR1 = Irrigated (D = 20 mm)

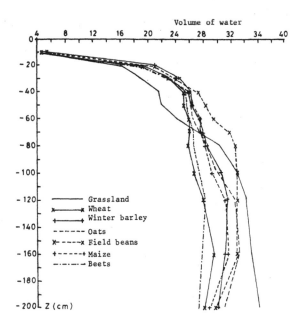

Fig. 11   Water limit profiles (withering) for the various crops

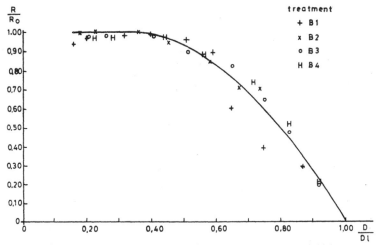

Fig. 12   Relationship between productivity and water deficit

Finally, empirical models have been developed. They are the following:

- An evaporation model for bare soil (Dautrebande et al., 1984)
- Real evapotranspiration model for alluvial grasslands (Ben Harrath et al., 1985)

47

- A model describing the evolution of the relative yield of grasslands for every level of nitrogen fertilizer as a function of the relative water deficit (Figure 12) (Ledieu et al., 1985)

2.2. Areas for further investigation

A general model describing the effects of drainage and irrigation should be constructed. On the experimental level, it is necessary to expand our knowledge of soil drying in the upper layers as a function of root development and cropping practices, particularly in the germination stage and in the initial phase of crop growth.

3. CONCLUSIONS

The drainage and irrigation studies conducted over twenty years have provided answers to a number of scientific and technical problems. But they also raise new questions that require further research and experimentation. The utility of mathematical models for the purpose of data synthesis and statistical studies in relation to climatic variability has been demonstrated.

REFERENCES

Ben Harrath, A., Frankinet, M. and Ledieu, J. 1977. Indice de la sécheresse de 1976 sur les conditions d'humidité du sol et les rendements agricoles. Bulletin de la Société Belge de Pédologie XXVII,2: 207-224.
Ben Harrath, A., Nyskens, J., Ledieu, J., Dautremande-Gaspar, S. and Sine, L. 1981. Rentabilité de l'irrigation de complément sur prairies en Belgique et choix du matériel. Revue de l'Agriculture 5.
Ben Harrath, A., Ledieu, J., Dautrebande-Gaspar, S. and Noirfalise, A. 1985. Modélisation de l'évapotranspiration réelle de végétations herbacées. Agric. Water Manag. 10: 1-13.
Calembert, J. and Sine, L. 1962. Evaluation de la conductivité hydraulique des sols en place. Pédologie XII: 129-143.
Dautrebande-Gaspar, S., Ledieu, J., Ben Harrath, A. and Frankinet, M. 1983. Modélisation de l'évaporation d'un sol nu. Bull. Rech. Agron. Gembloux 18(3): 189-196.
Rapports IRSIA 1960 to 1983. Génie Rural II, Faculté des Sciences Agronomiques de l'Etat, Gembloux.
Sine, L. and Vermeiren, L. 1965. Contribution à l'étude hydraulique des nappes de drainage - 2è partie: régime variable (suite). Bull. Inst. Agr. et Stat. Rech. Gembloux 33.3: 424-440.
Sine, L. and Dautrebande-Gaspar, S. 1978. Drainage dans les sols stratifiés: élaboration d'un abaque. Annales des Travaux Publics de Belgique 4.
Sine, L., Ben Harrath, A. and Ledieu, J. 1979. Interaction entre fumure azotée - Irrigation de complément. Séminaire Mouvements de l'eau et efficacité des apports d'eau. Bologna, 6-9 November 1979.

Irrigation demand and utilization of irrigation water on sandy soils in Denmark

P.C.THOMSEN
Institute of Irrigation & Agricultural Meteorology, Jyndevad, Tinglev, Denmark

ABSTRACT

About 50% of agricultural land in Denmark is composed of sandy soils. Therefore supplementary irrigation is important to maintain a high and steady production. Good returns on irrigation equipment and efficient utilization of water resources are determined by adapting crop, irrigation capacity and irrigation strategy, mainly based on knowledge of how the crops react on water stress in different growth stages.

Experiments are being carried out in which the crop is exposed to different degrees of water stress during the growth period. Crops under investigation are grass, winter wheat, winter barley, rye, perennial ryegrass for seed, rape, peas, potatoes, maize, fodder beet and poppy. The influence of supplemental irrigation on other growth factors is also being investigated in these experiments, and emphasis has been put on the interrelationship between irrigation and nitrogen. An understanding of these interrelations may lead to a more efficient use of water resources and irrigation capacity and of other growth factors. E.g. nitrogen was better used by the crops, resulting in greater efficiency in the use, thereby reducing the risk of undesirable effects on the environment.

During the next couple of years a system for irrigation scheduling will be designed, using the results from the experiments and the water balances for crops or fields. This will give farmers a greater opportunity for better irrigation scheduling, assisted by modern computer techniques.

1. INTRODUCTION

About 67% or 2.9 mill. ha of land in Denmark is agricultural land. In 50% of the land the maximum amount of plant available water varies between 60 and 80 mm.

The difference between precipitation and potential evapotranspiration in May and June was 63 mm on the average over the period 1965-1984. It varied between a surplus of 23 mm and a deficit of 144 mm. A large deficit in this important part of the growth period may lead to considerable reductions in yields on soils with a low amount of plant available water. Supplementary irrigation is therefore necessary in order to obtain a high and stable production on sandy soils.

In the sixties and in the beginning of the seventies livestock was held all over the country, on both clay and sandy soils. However the last 10 to 15 years livestock has been concentrated mainly on sandy soils in Jutland and the number of cattle per hectare has increased. The demand for stability

and quality in fodder crop production has therefore increased, giving rise
to greater investments in irrigation equipment, because the area under irrigation expanded. Changes in farming practices have given better returns on
investments in irrigation equipment, even on farms without fodder crops.

On a farm there may be fodder crops as well as commercial crops grown.
To obtain an efficient use of the given water resources, it is necessary to
know the reaction to water stress of the different crops during the growth
period. The demand for irrigation scheduling is therefore great.

2. SOILS

The agricultural land in Denmark is composed mainly of deposits from
glacial periods - partly moraine and partly from glacial streams. The country was not entirely covered with ice during the last glacial period. This
has not only caused a great variation in soils between regions, but also
within a region. Consequently, sandy soils account for a different percentage of the agricultural land in different areas (table 1, figure 1). The

Fig. 1   The position of the counties in Denmark

TABLE 1   Percentage sandy soils in relation to the agricultural land in Denmark and in the different counties, and the agricultural land and the arable land in relation to the total area (after Danmarks Statistik, 1983; Ministry of Agriculture, 1985)

Particle size	Soil class			Agricul-tural land	Arable land
	Coarse sandy	Fine sandy	Loamy sand		
<2 µm, %	0- 5	0- 5	5-10		
2 µm- 20 µm, %	0-10	0- 10	0-15		
20 µm-200 µm, %	0-50	50-100	0-40		
Locality	%	%	%	%	%
Denmark	25.1	6.1	28.1	67	61
Bornholm					
Bornholm	2.2	0.3	6.8	62	61
Sjaelland					
København	0.0	0.0	18.2	20	19
Frederiksborg	1.0	2.7	67.5	48	44
Roskilde	0.4	0.3	12.1	64	62
Vestsjaelland	1.4	2.1	24.2	70	66
Storstrøm	1.5	1.6	12.4	73	71
Fyn					
Fyn	2.2	1.7	34.7	71	67
Jylland					
Arhus	17.3	4.0	44.1	64	59
Vejle	25.6	0.0	29.4	68	62
Nordjylland	21.8	36.8	25.5	68	59
Viborg	26.6	12.8	37.1	66	58
Ringkøbing	59.5	7.6	18.0	66	61
Ribe	53.5	1.9	31.6	67	58
Sønderjylland	40.7	0.4	19.5	74	70

plant available water in coarse sandy soils is about 60 mm, in fine sandy soils about 80 mm and in loamy sandy soils between 80 and 110 mm.

In the southwestern part of the country the soils are mainly coarse sandy, in the northern part of Jutland mainly fine sandy, while in the eastern part of Jutland and the eastern islands the soils have a good available soil water capacity. The distribution of soil types within a county has no influence on the percentage of agricultural land. However, in areas with sandy soil a smaller percentage of the land is used for arable crops.

3. CLIMATE

Denmark has a temperate, humid coastal climate. On the average the growth period starts on 1 April. The most drought sensitive periods for most of the cultivated crops are in May and June. In these months potential eva-

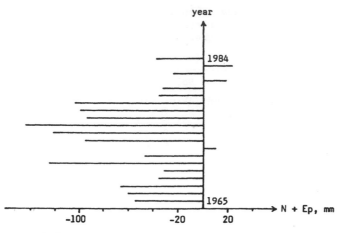

Fig. 2   Precipitation + evapotranspiration, 1965-1984 (after Medd. Statens Planteavlsforsøg)

TABLE 2   Precipitation, N (mm), and potential evapotranspiration, E_p (mm), in provinces. Averages for the period 1965-1984 (after Medd. Statens Planteavlsforsøg)

Province	May-June			April-October		December-November
	N	E_p	$N-E_p$	N	E_p	N
Bornholm	79	167	-88	359	482	642
Sjaelland	99	174	-75	371	502	615
Fyn	103	150	-47	390	408	662
Østjylland	100	166	-66	393	476	673
Nordjylland	108	162	-54	420	459	719
Midtjylland	116	160	-44	477	461	835
Vestjylland	109	169	-60	474	496	844
Sydjylland	112	169	-57	475	478	804

TABLE 3   Area under irrigation in hectares and as percent of the agricultural land, and number of farms with irrigation equipment (after Danmarks Statistik, 1983)

Year	Irrigated area		Number of farms
	ha	%	
1972	130 000	4.5	5 200
1976	201 000	6.9	8 800
1980	391 000	13.5	15 400
1982	390 000	13.4	15 000

TABLE 4   Area under irrigation in hectares and as percent of the agricultural land, number of cows and coarse sandy and fine sandy agricultural area in hectares (after Danmarks Statistik, 1983; Ministry of Agriculture, 1985)

	Irrigated area		Number of cows	Coarse sandy	Fine sandy
	1000 ha	%	1000	1000 ha	
Denmark	390	13.5	1059	684	285
Bornholm					
Bornholm	0.5	1.5	8.5	0.8	0.1
Sjaelland					
København	0.8	7.8	0.6	0.0	0.0
Frederiksborg	5.5	8.7	10.9	0.6	1.8
Roskilde	1.9	3.2	6.6	0.2	0.2
Vestsjaelland	10.6	5.1	36.7	2.9	4.4
Storstrøm	7.8	3.1	32.3	3.7	4.0
Fyn					
Fyn	11.9	4.8	78.4	5.5	4.2
Jylland					
Arhus	19.3	6.6	88.1	50.9	11.8
Vejle	31.0	15.1	78.1	52.4	0.0
Nordjylland	28.4	6.8	184.0	58.9	185.4
Viborg	24.0	8.8	131.3	72.9	35.1
Ringkøbing	107.7	33.4	145.6	191.6	24.5
Ribe	85.5	40.7	124.6	112.4	4.8
Sønderjylland	55.0	18.9	133.5	118.0	1.2

potranspiration, E_p, is usually greater than precipitation, N (table 2).

The deficit is smallest in the areas where sandy soils dominate. There the average deficit in May and June has the same magnitude as the available water capacity of coarse sandy soils. For the country as a whole, there is a great variation in the average figures ranging from a surplus of 23 mm to a deficit of 144 mm (figure 2).

The magnitude of precipitation and evapotranspiration in short periods within the growth period is of great importance for plant production. One week without precipitation may cause a deficit, which limits plant growth due to water stress.

4. IRRIGATION CAPACITY

The area of agricultural land under irrigation nearly tripled over the period 1972 to 1982 (table 3).

In 1972 irrigation with fixed sprinklers dominated, while today irrigation machines are the most common used system. Irrigation systems are dominant in areas with sandy soils and high number of cattle per ha (table 4). In other areas there is a considerable market gardening with irrigation systems.

The source of irrigation water is mainly groundwater. Groundwater is easiest to obtain in the southwestern part of Denmark, where irrigation demand is the highest. Pumping of groundwater is restricted and permission is required. Permission is given to pump a certain amount of water per year for a limited number of years.

5. IRRIGATION COSTS

Irrigation costs depend upon the investment. The depth of the groundwater well and the distance to and between the fields play a role too. An investment of DKr 7000 (DM 1920) per ha is a realistic average, which results in a fixed cost of DKr 1170 (DM 320) per ha. Running costs result in a cost of about DKr 4.80 (DM 1.3) per ha per mm. With an average irrigation demand of 100 mm the running costs are DKr 4800 (DM 1300) per ha per year.

6. YIELD INCREASE WITH IRRIGATION

Irrigation experiments have been carried out on coarse sandy soil over the last fourty years at the Government Research Station, Jyndevad. Yield increases for different crops cannot be immediately compared, as the investigations were conducted in periods with different climatic conditions. Taking this into consideration and allowing for loss by handling and storage, net yields with and without irrigation can be calculated for different crops (table 5). On fine sandy soils the available water capacity is about 20 mm greater than on coarse sandy soil. Therefore the demand for irrigation water is lower resulting in smaller yield increases. The returns of irrigation for different crops on coarse sandy and fine sandy soils, calculated on the basis of yield data in table 5, crop price and total costs, are given in table 6.

The returns of irrigation of different crops vary considerably. Under irrigated conditions the yield of most crops is comparable to yields on soils where water does not restrict plant growth. Therefore irrigation enables more crops to be cultivated on sandy soils. The average economic yield increase is negative for some crops on fine sandy soils and is always lower than that on coarse sandy soils (table 6). The variation in yields

TABLE 5 Average yields and yield increases caused by irrigation per ha on coarse sandy soils and fine sandy soils. Based on results from experiments at the Government Research Station, Jyndevad

Crop	Yield	Yield increase %		Applied water, mm	Yield increase per mm
Coarse sandy soils					
Spring barley, hkg	31	12	42	77	16 kg
Winter barley, hkg	45	11	24	77	14 kg
Oat, hkg	30	13	47	77	17 kg
Rye, hkg	38	9	24	65	14 kg
Winter wheat, hkg	37	20	54	85	24 kg
Rape, hkg	18	8	53	77	10 kg
Peas, hkg	35	15	43	77	19 kg
Potatoes, hkg	420	90	21	68	132 kg
Fodder beets, FU	9 300	1900	20	83	23 FU
Grass forage, FU	5 700	1900	33	152	13 FU
Fine sandy soils					
Spring barley, hkg	37	6	16	40	15 kg
Rape, hkg	21	5	24	40	6 kg
Potatoes, hkg	415	55	13	50	26 kg
Fodder beets, FU	10 000	1200	12	45	27 FU
Grass forage, FU	6 100	1500	25	120	13 FU

with no irrigation are considerably lower on fine sandy soils than on coarse sandy soils. Therefore investment in irrigation equipment for coarse sandy soils is most valuable when stable fodder crop production is important or when important commercial crops, e.g. potatoes, are grown.

The type of crops under cultivation, with and without irrigation, depends among other things on the average economic yield and its variation. Crops with high yield increases with irrigation will naturally give low yields and a large variation in yield over the years, without. Irrigation on cattle farms stabilizes fodder crop production thereby avoiding large surpluses or deficits. In addition the quality of the crop is improved. The area of land, designated for the necessary fodder crops, can be reduced through irrigation thereby freeing land for other crops. Mixtures of grass and clover become more attractive with irrigation, resulting in savings in nitrogen fertilizer, but a certain production requires a larger area than production of pure grass swards at a high N-level. At the moment alternatives for fodder crops are economically realistic, and good economic yields of commercial crops are attainable. These crops are therefore also under cultivation on livestock farms.

TABLE 6   Irrigation costs and returns for different crops on coarse
sandy and fine sandy soils. Based on results from experiments at the
Government Research Station, Jyndevad (after COWICONSULT, 1985)

Crop	Irrigation costs	Returns	
	DKr per ha	DKr per ha	DKr per ha per mm
Coarse sandy soils			
Spring barley	1.540	318	4.1
Winter barley	1.540	163	2.1
Oat	1.540	459	6.0
Rye	1.482	206	+3.2
Winter wheat	1.578	1.618	19.0
Rape	1.540	1.423	18.5
Peas	1.540	1.646	21.4
Potatoes	1.496	2.104	30.9
Fine sandy soils			
Spring barley	1.362	433	10.8
Rape	1.458	491	12.3
Potatoes	1.362	702	11.7

On coarse sandy soils irrigation is an important factor in obtaining a
stable yield of good quality and also contributes to an efficient use of
other growth factors, especially nitrogen.

7. IRRIGATION RESEARCH

The need to acquire an insight in the relationship between irrigation
and other growth factors leads to investigations at the Government Research
Station, Jyndevad. An example of this type of investigation forms field experiments with irrigation intensity, nitrogen fertilization and cutting
frequency in perennial ryegrass. Each of these factors has a great influence
on grass yield and quality, and the effect of one parameter is dependent on
the level of the others. Better understanding of these interrelations may
lead to a more efficient use of irrigation water and nitrogen fertilizer.

The purpose of another experiment is to optimize N-application on spring
barley under irrigation. N-mineralisation, N-uptake and leaching during the
first part of the growth period are measured. Results from this experiment
are expected to be a basis for advising on fertilizing strategies, which
minimize N-leaching and maximize the efficiency of N-uptake.

Knowledge of the reaction of different crops to water stress is necessary
to optimize the use of irrigation water. Investigations into how crops react
upon different water deficits in different growth stages has been an impor-

tant subject over the past couple of years. The purpose is to discover when drought sensitivity is the greatest for different crops and which crop responds with the highest yield increase to irrigation at a certain time.

An experiment on spring rape seed was carried out according to the following plan:

Tension (mbar) 22 cm below surface or deficit in mm in the root zone at the time of irrigation

Treatment	Phase 1	Phase 2	Phase 3
1	non-irrigation	non-irrigation	non-irrigation
2	40 mm	450 mb	450 mb
3	40 mm	450 mb	40 mm
4	40 mm	40 mm	450 mm
5	40 mm	40 mm	40 mm
6	850 mb	850 mb	850 mb

Phase 1: from sowing to 20 May
Phase 2: from 21 May to start of flowering
Phase 3: from start of flowering until harvest

Time of irrigation is determined by tensiometers that are able to indicate soil water potentials above 1 bar. At lower potentials the deficit is determined by neutron scattering.

Results of this experiment for the year 1980 are given in table 7. During that year there was only an irrigation demand in phase 2 of table 7. Equal yield increases were obtained with one irrigation at 40 mm deficit, 2 irrigations at 30 mm deficit and 3 irrigations at 20 mm deficit. It should be noted that 40 mm is about 2/3 of the total available water in the soil. The results indicate that spring rape seed is fairly drought resistant in the period before flowering. The drought sensitivity of a crop can be explained by conducting this type of investigation over a couple of years with differing climatic conditions.

Irrigation water demand and drought sensitivity have been investigated in the following crops: potato, spring barley, winter barley, winter wheat, rye, spring rape seed, winter rape seed, pure grass, mixture of grass and clover, peas, maize, fodderbeets, perennial ryegrass for seed and poppy. A few years ago many of these crops were not grown on coarse sandy soils, but with proper irrigation scheduling high and stable yields have been made possible for all crops.

TABLE 7   Yield without irrigation and yield increase caused by irrigation of spring rape in 1980 on a coarse sandy soil

| Treatment | Irrigation, mm | | | | Yield and yield increase |
	Phase 1	Phase 2	Phase 3	Sum	hkg per ha
1	0	0	0	0	22.5
2	0	3x20	0	60	4.3
4	0	1x40	0	40	4.0
6	0	2x30	0	60	3.9
LSD_{95}					1.5

In a research project the possibility to increase the capacity of plant
available soil water and to affect plant reaction to drought stress is under
investigation. The effect on drought stress, water uptake and yields of
crops from deep loosening of the soil organic matter amendment, potassium
fertilizing and previous crop is being investigated.

Investigations into the drought resistance of cultivars and the improvement of meteorological forecasts, to support irrigation scheduling, may improve efficiency in use of irrigation water.

8. IRRIGATION SCHEDULING

One of the purposes of the experiments, which have been carried out at
the Government Research Station, Jyndevad, is to quantify K in the relation:

$$1 - Y_a/Y_p = K(1 - E_a/E_p)$$

for different crops and in different growth periods. In this equation Y_a is
the yield at the actual soil water status, Y_p the yield when water does not
restrict plant growth, E_a is the actual evapotranspiration and E_p the
evapotranspiration at field capacity. During the next couple of years a
system for irrigation scheduling will be initiated on a data medium. In this
way farmers who have terminal facilities can easily make use of results from
irrigation experiments in scheduling irrigation. The knowledge of K-values
are then of great importance as basis for decision of irrigation of a particular crop at a particular time.

The climate determined evapotranspiration is obtained in 39 evaporation
pans distributed throughout the country. Weekly values from these pans are
used by the farmer together with their own precipitation data and results
from irrigation experiments to determine the amount and time of irrigation
for different crops and fields.

9. DISCUSSION

In the southwestern part of Denmark a considerable part of the agricultural land is coarse sandy soils. In one county even 60% is coarse sandy. A low available soil water capacity in these soils causes great variations in yields if there is no possibility for supplemental irrigation. Where supplemental irrigation is possible yield and variation in yield are comparable with crop yields on soils with good water holding capacities.

The average yields in table 5 are based on the fact that water resources for irrigation are sufficient, even in dry years. Permission to pump water for irrigation is usually given for an average irrigation demand, often 1000 m^3 per ha. In exceptionally dry years irrigation demand can exceed 2000 m^3 per ha. In these years it is difficult to obtain permission to pump more water because of the consequences of groundwater pumping for river flow. In the future it is expected that it will become even more difficult to obtain permission to pump water for irrigation.

With an insufficient irrigation capacity optimum yield cannot be achieved for all crops. When the average yield increases are only 50% of the reported ones irrigation is still profitable on some crops (table 8).

TABLE 8   Percentage of the average yield increase, which causes zero returns of irrigation on coarse sandy soils (after COWICONSULT, 1985). For yield increases see table 5

Crop	%
Spring barley	84
Winter barley	91
Oat	78
Rye	108
Winter wheat	52
Rape	54
Peas	54
Potatoes	53

With knowledge of the reaction of the crops to drought conditions in any part of the growth period, the yield reduction caused by limited water resources can be minimized. However, there will always be an uncertainty about when to use the limited resources because the climate in the remaining part of the growth period is unknown.

In Sønderjylland about 40% of the agricultural land is coarse sandy. Less than half of this area is under irrigation. It is therefore profitable to double the irrigation capacity in Sønderjylland. With permission for 1000 m^3 per ha per year the demand for extra irrigation water is 50 million m^3. The value to society of higher yields due to irrigation on coarse sandy soil can be calculated at DKr 1.25 per m^3 or about DKr 1250 (DM 340) per ha per year, with the current prices of crops (COWICONSULT, 1985). There will also be an increased earning of DKr 2500 (DM 685) per ha per year in foreign currency and an increase in employment of about 4 hours per ha per year (COWICONSULT, 1985). These economic consequences have to be seen in relation to other needs for water resources, and the consequences to the environment by increasing the quantities of water pumped from groundwater.

In addition to yield increases, supplemental irrigation also contributes to an efficient use of other growth factors. Consequently negative influences on the environment may be reduced. Investigations into irrigation in relation to other growth factors and the environment will be continued.

REFERENCES

COWICONSULT. 1985. Markvanding i Sønderjylland. Produktionsøkonomiske beregninger. Sønderjyllands Amtskommune. Rapport. 74 pp.
Danmarks statistik. 1983. Agricultural statistics 1982.
Ministry of Agriculture. 1985. Arealdatakontoret. 1. Afdeling 6. Kontor.
Medd. Statens Planteavlsforsøg. Annual reports on precipitation, evaporation and water balance, 1966-1985.

Effect of irrigation on agriculture: Water availability and crop yield

M.E.VENEZIAN SCARASCIA
Ministero Agricoltura e Foreste, Bari, Italy

ABSTRACT

In South Italy the main factor constraining crop yield is the soil water shortage. In spite of adequate values of annual precipitation and soil water storage capacity the high levels of incident radiation and wind speed cause a massive evaporation flux. Under these environmental conditions the stage at which grain crops are susceptible to drought appear to be critical coincides with the period of maximum atmospheric evaporation demand.

Due to plant water status the rates of leaf expansion and plant growth largely determine the crop yield. Therefore meteorological measurements supported with physiological data provide one of the best means to establish the time of plant water stress: in particular the infrared thermometry along with the strategy of limited irrigation allow to increase the water use efficiency (WUE) of the crops.

1. INTRODUCTION

Irrigation is an agricultural practice to supplement a deficiency of the climate, the imbalance between the water supplied by precipitation and the evaporative demand of the atmosphere. During the period of precipitation the soil stores a portion of the supplied water. When the rain ceases evapotranspiration reduces the stored water. If the rate of evapotranspiration is high and the soil water content reaches a critical level for plant growth, the need of irrigation arises. Therefore the impact of climate on the water balance of crops and thus on irrigation demands depends upon the physical properties of the soil – chiefly the soil water storage capacity – and the physiological characteristics of the crop. In broad terms it can be stated that the irrigation requirement is determined by the type of soil, climate and type of the crop (Figure 1).

2. THE CLIMATE

The Italian peninsula is located between 36° and 47°N, its annual temperature being between 11° and 19°C. It is characterized by considerable differences in total precipitation as well as in rainfall distribution. The Mediterranean climate has mild winters with maximum precipitation and hot and rainless summers.

From North to South the precipitation amount is decreasing and meanwhile

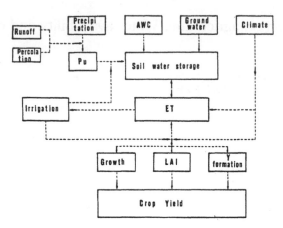

Fig. 1   Diagram of the components of the soil water – crop – atmosphere
system

solar radiation, temperature and evaporation are increasing. The mean annual precipitation ranges from an average of 1000 mm in the North – with a
maximum of 3000 mm in the Alps – to an average of 600 mm and less in the
South (Figure 2).

The geographic conditions of Italy cause precipitation patterns to
change from year to year and from region to region: its occurrence is un-
predictable; consequently the risk of crop failure and unsatisfactory
yields is high.

Because of lower rainfall and higher evapotranspiration rates during
the late spring, the soil water depletion is increasing in the central-
southern part of the peninsula. Therefore summer crops give little production without irrigation. Sometimes dryness occurs in early spring and in
this case the winter crops also experience water stress.

There appears to be a discrepancy between the available soil water, the
atmospheric demand and the peak period of water needs. The climate requires
a higher supply of water during the period of the highest physiological requirements for the crop (Figure 3).

Irrigation can be used to stabilize and increase crop production. Moreover in the South irrigation makes it possible to grow summer crops sown
either in early spring or, as a second crop, after winter cereals. Irrigated summer crops are very productive because high solar radiation in summer
brings about favourable conditions for photosynthesis provided suitable
soil water conditions are maintained.

3. IRRIGATED LAND

The area under irrigation has increased in Italy from 2150 million hectares in 1942 to 3345 million hectares in 1979. Recent increases have been especially large in central and southern Italy. Of the total irrigated land 69% is located in the northern area, 21% in the southern regions and 9% in the centre. In the North irrigation is applied to 29.7% of the agricultural land, in Central Italy to 7.2%, in the South to 6.7% and on the isles to 4.5%.

Irrigated land is located predominantly in the plains (72% of the total) and over 50% of the total agricultural production comes from irrigated land. The used systems are surface irrigation (47%) and sprinkler irrigation (24%).

Usually in irrigation fixed amounts of irrigation water at fixed intervals are applied with the tendency to ignore the rainfall amounts. In the irrigated areas sufficient water is available so rainfall is not generally considered as an essential part of the crop water requirement. Only when considerable amounts of rain occur irrigations are delayed.

The irrigation scheduling based on estimated daily evapotranspiration and soil depletion allows predicting of the optimal timing and amounts of irrigation taking into account previous and/or expected precipitation. For this purpose the irrigationist needs precise quantitative information on irrigation amounts and timing. Detailed investigation of climate, soil environment and physiological behaviour of the crop can provide the required informations.

4. CROP WATER REQUIREMENTS

Applying the quantity of water a crop requires for not limiting growth and yield formation often causes over-irrigation. There may be national and regional economic advantages in less water but then a real estimate of crop water requirement and the consequences of it have to be known.

The rate at which water is used by a crop is dependent on a number of factors which include species, variety, age, climate and soil water availability. Consequently the water use of a crop will vary from region to region and in some case from field to field. A common method to estimating crop water use is to calculate the evapotranspiration of a reference crop (ET_o) from climatic data and then to convert this to the actual crop evapotranspiration by means of a crop factor.

In general terms the crop water requirement is equivalent to the rate of evapotranspiration necessary to sustain optimum plant growth and yield formation. The accuracy of the determination of crop water requirement will

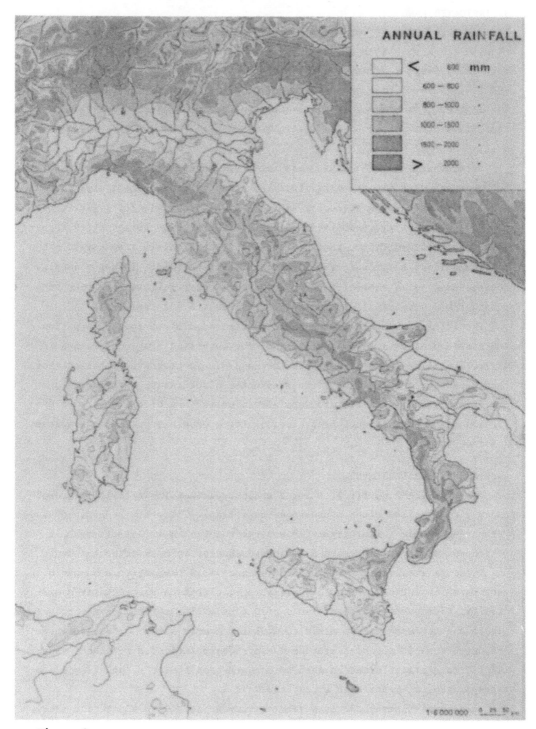

Figure 2

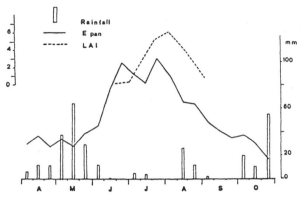

Fig. 3   Trends of rainfall, E_v and sunflower LAI in South Italy

be largely dependent on the type of method chosen to estimate the evapotranspiration. No single method to compute ET_o using meteorological data is
satisfactory for all climatic regions without local calibration (Burman et
al., 1983). For instance with pan evaporation (E_v), ET_o and E_v are governed
by the same meteorological factors. Consequently there is a strong correlation between them, but as open water and crop surface react differently to
the meteorological conditions it is necessary to establish experimentally
the relationship between ET_o and E_v. This relationship is generally obtained
from a linear regression or from lysimeter data using weekly or monthly
measurements of ET_o and E_v. The relation then is often given as a simple
ratio. This form is very convenient for computing irrigation requirements.
Reported values of the ratio ET_o/E_v range from 0.6 to 2.0, but most of the
variation can be explained by pan type and exposure of pan. According to
WMO the most commonly used evaporation pan is the USDA Weather Bureau class
A pan, which is easy to install and to maintain. To obtain reliable data it
is important to leave the water level in the pan not for long periods below
the minimum required level and to place the pan in the same conditions as
the crop to be irrigated.

Values of crop evapotranspiration (ET_c) based upon lysimetric measurements have been obtained in two Italian localities with very different environmental conditions: Padova in the North (Giardini et al., 1976) and
Foggia in the South (Rizzo et al., 1980). Comparing the evapotranspiration
rates of a grain maize crop with the values of ET_o estimated according the
method of Blaney-Criddle and with the evaporation from a class A pan (E_v),
the ET_c/E_v ratio showed a good fitting and a low variability in both cases.
This result permits to evaluate the ET_c of a maize crop under good growing
conditions only making use of the E_v values (Figure 4) and to ascertain the

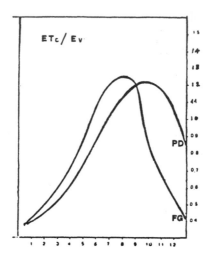

Fig. 4   Trends of ET_c/E_v ratio during corn crop cycle in North (PD) and
South Italy (FG)

'crop coefficients', i.e. the multiplication factor for soil water not limiting ET_c. The crop coefficient appeared to be very much the same for both
localities.

5. ENERGY BUDGET AND WATER BUDGET

Because the evaporation process is the connecting link between the water
budget and the energy budget it is possible to estimate its magnitude either
through the water balance equation or through the energy budget equation.

When water is readily available ET_o is the largest term of the energy
balance equation and for this condition it can be said that as a first approximation evaporation is proportional to net radiation. In an investigation
carried out for many years in the Southern region of Puglia it was observed
that the primary climatic parameters influencing evaporation in that area
are solar radiation and - for the summer period only - the daytime wind; the
SW and SE winds are associated during summer with the highest E_v rates. The
regression equation

$$E_v = -2.41 + 0.017 \, Rd + 0.007 \, W$$

is well fitting to measured data (Venezian Scarascia et al., 1982a).

According to the energy balance equation, factors that modify the net
radiation will also affect the evaporation; consequently the albedo that
affects the radiation balance provides useful information on the estimated
evaporation. Two agronomic parameters, namely leaf area index (LAI) and leaf

water potential, having a great influence on the crop yield, are correlated with albedo values. The albedo is higher with increasing LAI values and with decreasing leaf water content (Venezian Scarascia et al., 1982b).

To estimate evapotranspiration through the energy balance we need to investigate the net radiation, the ground flux density and the sensible heat flux. Particular difficulties can arise in estimating of the latter because the atmosphere is an open system. The air flow over the crop surface continuously brings new air in contact with the surface and mixes the air with the air stream at higher levels. Consequently the energy balance method cannot provide precise informations on evaporation flux.

In the South of Italy two methods were compared to measure ET_c of a sorghum crop, one based on lysimeter measurements and the other on estimated values of the Bowen ratio (Maracchi et al., 1979). The results showed that the evapotranspiration is larger than the net radiation for additional latent heat is provided by hot dry air flowing over the irrigated fields. This feature of the energy balance, the local advection of sensible heat, cannot be measured with the energy balance method.

6. CROP RESPONSE TO WATER

The volume of irrigation water should be closely related to evapotranspiration so as to replenish the water deficit throughout the root zone. However, the increasing scarcity of water and the growing food demand are pressing for a maximum water efficiency i.e. maximum crop yield per unit of water input.

The relationship between crop yield and water use has been studied by many workers (Vaux and Pruitt, 1983). Linear correlation has been developed between accumulated evapotranspiration and biomass production in wheat, sorghum, oats and alfalfa (Hanks et al., 1969).

Using mean crop growth rate ($\overline{CGR}$) values obtained from 16 harvests of 18 cvs of alfalfa recorded during a 3 years period, a linear relationship between $\overline{CGR}$ and E_v was observed; the $\overline{CGR}-E_v$ function for alfalfa grown in South Italy can be represented by (Venezian Scarascia and Losavio, 1982c):

$$\overline{CGR} = 1.049 + 0.65 \, E_v \quad (r = 0.81)$$

This relation is the simplest and the most suitable when total dry matter is the desired product. For grain crops the yield is not always proportional to ET: the functional relation between yield and the seasonal irrigation depth appears to be curvilinear (a second degree polynomial). In order to

apply economic maximization techniques to water production function knowledge of this relationship is required. This information is usually obtained through the regression analysis of field experimental data in which water is a variable of the treatments.

During the past 20 years the Italian scientific community (university,
CNR, MAF) has conducted many cooperative trials on crop response to increasing irrigation water with a unified methodology. The irrigation scheduling
was applying a constant volume of irrigation water to replenish a fraction
of the total evapotranspired water, estimated on the basis of pan A evaporation. Hence the increasing total irrigation water volumes were obtained
applying irrigation water with increasing frequency. Cavazza (1975) analysing the findings of the early investigations, noted that Mitscherlich's
first approximation model can be considered valid because the mean curves
resulting from the trials conducted for many years fitted this model. Moreover through the same analysis he was able to determine the large variability of the 'maximum potential yield' for the same site and to estimate
rainfall efficiency.

By means of the same model Quaglietta Chiarandà et al. (1982) studied
in six sites of central, southern and insular Italy during a three years
period the water requirements of grain maize. The curve fitted with high
significancy to the observed data (Figure 5). It was found that the potential yield differs between sites and over years, while the 'action coefficient of water' differs between sites only.

In a series of trials, designed with the aim to determine the water re-

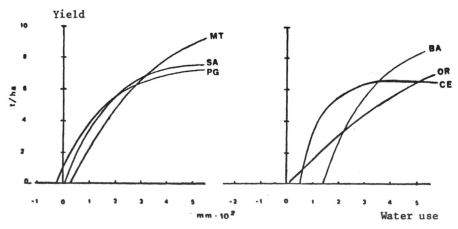

Fig. 5   Maize grain yield-water use relationship in six Italian sites

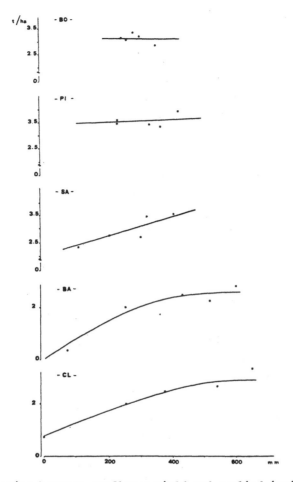

Fig. 6   Relation between sunflower yield and applied irrigation water

quirement of sunflower grown under optimal and suboptimal conditions of water availability in 5 Italian sites, Venezian Scarascia et al. (1981) found
a curvilinear relationship between crop yield and water applied for the
Puglia and Sicily sites and a linear relation for the trial conducted near
Salerno. Data from the experimental fields of Bologna and Pisa showed that
irrigation was not necessary in these sites due to high rainfall during the
growing season (Figure 6).

The single relation accounting for seasonal water use only may be sufficient for grain crops because other results are probable when irrigation
is based on growth stage.

7. PHYSIOLOGICAL APPROACH

An important factor in water management for grain crops is the occurrence of soil water stress imposed on the crop in relation to its physiological stage. Many studies have clearly demonstrated that the stress sensitivity is generally greatest in the floral through pollination and in the first part of the grain filling stage. Withholding irrigation during the early vegetative stage reduced the grain yield of sunflower by 4%. The yield was reduced by 17% imposing stress during the active growth stage. The reduction was sharply increasing to 76% if the irrigation was withheld during the anthesis stage (Figure 7) (Venezian Scarascia et al., 1981).

Early water stress causes a reduction of the assimilatory surface area. Among the effects of the stress conditions reduced cell enlargement is the most prominent. The leaf area index (LAI) of grain sorghum is reduced proportionally to the amounts of used water, that is

$$LAI_{max} = 0.93 + 0.0116 \ ET_c$$

The lower LAI affects the mean CGR and the biomass production negatively. The crop intercepts less solar radiation so that the synthesis of the assimilates decreases and the grain yield is reduced (Venezian Scarascia et al., 1983).

Denmead and Shaw (1960) suggested that stress imposed in one stage hardens the plant against damage from stress in a later stage. Stewart et al. (1975) termed this effect a 'conditioning factor' since the reduction in plant size, caused by early water stress, appears to condition the crop so

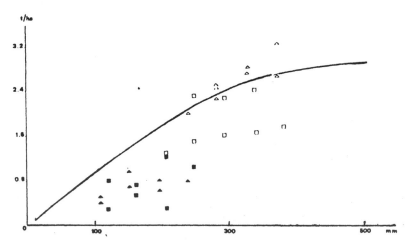

Fig. 7 Maize grain yield affected by water deficit during flowering stage (■▲) deviating from the Mitscherlich curve

70

that a water deficit during more critical later stages had less effect on the yield.

Where water is a limiting resource the objective of the irrigation management may shift from obtaining maximum yield to obtaining maximum economic production per unit of applied water. For this aim the crop water deficit may not be totally avoided but instead favourably 'controlled'. It appears that water stress regardless of timing tended to reduce the water use efficiency, the reduction being smallest when stress increases gradually throughout the growing season (Garrity et al., 1982).

For an economically efficient water management one may assume a certain physiological state as a criterion for the irrigation timing. The use of foliage – air temperature differences as a timing method for irrigation is based on the concept that water stress will cause a rise in plant temperature as reduced transpiration dissipates less of incoming solar radiation (Gates, 1976).

Idso et al. (1981) developed a plant stress index methodology to indicate water stress in using infrared thermometry for estimating the timing for water application. Plant water stress index values as timing criteria are not yet well-developed. However, in Minnesota irrigation scheduling on the base of crop temperature caused little or no reduction in yield (Stegman, 1983). Infrared thermometry along with a microcomputer system can offer a vehicle for a more rapid transfer of sophisticated irrigation scheduling technology in practice.

REFERENCES

Burmann, R.D., Cuenca, R.H. and Weiss, A. 1983. Techniques for estimating irrigation water requirements. Adv. Irr. 2: 335-394.

Cavazza, L. 1975. Influenza del volume stagionale di irrigazione sulla resa di alcune colture nel Mezzogiorno d'Italia. CASMEZ 52: 168-246.

Denmead, O.T. and Shaw, R.H. 1960. The effects of soil moisture stress at different stages of growth of the development and yield of corn. Agron. J. 52: 272-274.

Garrity, D.P., Watts, D.G., Sullivan, C.Y. and Gillery, J.R. 1982. Moisture deficits and grain sorghum performance: ET-Y relationship. Agron. J. 74: 815-820.

Gates, D.M. 1976. Water and plant life. Springer Verlag, Berlin, New York: 137-147.

Giardini, L. and Giovanardi, R. 1976. Consumo idrico del mais e contronto fra metodi per la stima della ET. Riv. Agron. 10: 187-201.

Hanks, R.J., Gardner, H.R. and Florian, R.L. 1969. Plant growth - ET relations for several crops in the Central Great Plains. Agron. J. 61: 30-34.

Idso, S.B., Jackson, R.D., Pinter, P.J., Reginato, R.J. and Hatfield, J.L. 1981. Normalizing the stress-degree-day parameter for environmental variability. Agric. Meteorol. 24: 45-55.

Maracchi, G.P., Raschi, A., Venerian, M.E. and Rizzo, V. 1979. Bilancio idrico di una coltura di sorgo. Atti Con. Naz. AIGR: 389-405.

Quaglietta Chiaranda', F. et al. 1982. Effetto di volumi irrigui crescenti sulla produzione di un ibrido di mais da granella in semina estiva in diverse località del centro-sud. Ann. Ist. Sper. Cer. 13, sup. 1: 95-120.

Rizzo, V., Di Bari, V. and Losavio, N. 1980. I consumi idrici per ET del mais da granella in coltura principale nell'ambiente del Tavoliere pugliese. Riv. Agron. 16: 263-274.

Stegmann, E.C. 1983. Irrigation scheduling: applied timing criteria. Adv. Irr. 2: 1-30.

Stewart, J.I., Misra, R.D., Pruitt, W.O. and Hagan, R.M. 1975. Irrigation corn and grain sorghum with a deficient water supply. Trans. ASAE 18: 270-280.

Venezian Scarascia, M.E. et al. 1981. The yield response of sunflower crop to increasing depths of irrigation and to drought periods in five Italian environmental conditions. 4° Cons. FAO on ECN sunflower: 113-130.

Venezian Scarascia, M.E., Castrignano', A.M., Losavio, N. and Mastrotilli, M. 1982a. La domanda evaporativa dell'atmosfera nella Murgia barese. Ann. Ist. Sper. Agron. 13: 157-186.

Venezian Scarascia, M.E., Maracchi, G.P. and Paloscia, S. 1982b. Stima delle caratteristiche delle colture agrarie mediante l'albedo. Ann. Ist. Sper. Agron. 13: 217-233.

Venezian Scarascia, M.E. and Losavio, N. 1982c. Crop growth rate of alfalfa in the mediterranean area. EGF Int. Congr., Reading.

Venezian Scarascia, M.E. et al. 1983. Dinamica dell'accrescimento a sviluppo di una coltura di sorgo da granella allevato in differenti condizioni di disponibilità idriche. Riv. Agron. 17: 382-391.

Discussion Session 2.1

Question from Mr. J. Feyen to Mr. D. Xanthoulis

In your paper you describe one empirical model for the relative yield
of pasture as a function of relative water deficit for several levels
of nitrogen application (see Figure 12 of your paper). Can you specify
with which production level R_o correspond? Is this the yield measured
on the plot with the highest N-application? If so, I am very doubtful
about the exactness of the given relationships.

Secondly, the model contains sections for modeling the water and
nitrogen balance. The relationship presented in Figure 12 is just giving the crop response.

Answer of Mr. Xanthoulis

R_o is the yield when the water deficit is equal to zero and the yield
is not measured on the plot with the highest N-application. This model
can also compute ETR, but not the nitrogen balance. Using the potential
evapotranspiration calculated from standard meteorological data, soil
moisture at field capacity and initial conditions of water content of
soil, the model can compute the productivity as a function of water
deficit for different nitrogen fertilizer levels if no other limiting
factors exist.

Question from Mr. A.L.M. van Wijk to Mr. D. Xanthoulis

You posed in your paper that there is no difference in water consumption of grasslands between different nitrogen fertilizer levels. But a
higher N-level goes with a higher dry matter production and thus with
a greater water consumption, unless the water use efficiency of grass
is greater at higher N-levels.

Answer of Mr. Xanthoulis

The plots were irrigated when the potential water deficit was 20 mm
for the two levels of nitrogen fertilizer. We observed that practically there was no significant difference in the evolution of water
storage in the soil between the two treatments. The same result was
obtained for the irrigated treatment when a water deficit of 60 mm was
allowed.

Comment of Mr. P.E. Rijtema to this answer

In Figure 12 of the paper of Prof. Dautrebande along both axes relative
values are given. This means that at a lower level of N-fertilization

the maximum water use at that level is less than at a higher N-level.
I suppose that this fact is the main reason that all points nearly fit
around a single curve.

Questions from Mr. P.J.T. van Bakel to Mr. P.C. Thomsen

1) Did you observe any effect of extracting groundwater on adjacent
non-irrigated parcels (via lowering of groundwater level)?

2) Did you consider possibilities of irrigation from surface water?

Answers of Mr. Thomsen

1) There was no effect of lowering groundwater tables because there
were no capillary rise possibilities, because of too coarse sand.

2) In future irrigation from surface water will be totally restricted,
because the amount of water in streams is not sufficient.

Question from Mr. E.A. Garwood to Mr. P.C. Thomsen

Would Mr. Thomsen care to comment on the absence of grass in Table 6
of his paper. Irrigation costs and returns for all other crops are
shown in Table 5 but not for grass?

Answer of Mr. Thomsen

For a fodder crop like grass there is no market price. The value of the
crop is individual for each farm. A general return for irrigation can
therefore not be given. Besides the stabilizing effect of irrigation
on the production of fodder crops throughout time is of great importance.

Question from Mr. P. Bazzoffi to Mrs. Venezian-Scarascia

Do you think that infrared thermometry can substitute classic methods
for water stress assessment in Italy like in Minnesota?

Answer of Mrs. Venezian-Scarascia

We are trying to set up a schedule for obtaining infrared thermometry
images and I am sure that it will be the best method for the southern
regions of Italy.

Drainage of clay lands for grassland production: Problems, effects, benefits

A.C.ARMSTRONG
Field Drainage Experimental Unit, ADAS, MAFF, UK

ABSTRACT

Although there are few technical problems with the drainage of grassland, the adoption of drainage on clay lands under grass is limited. This paper records current work, which aims to establish not only the direct effects of drainage, but also to evaluate them in an economic context.

1. INTRODUCTION

Grassland is an important component of British agriculture. In 1981, 60% (6.9 million ha) of the agricultural area of England and Wales was in grass (short or long term or rough grazing), of which 4 million ha (some 35% of the total area) was in long term permanent grassland (MAFF, 1982a). Such pastures predominate in the western half of the country, in areas of higher rainfall, and frequently on clay soils. Despite these environmental factors, only a small proportion of the drainage activity in England and Wales is undertaken in grassland (Armstrong, 1981), and the major areas of drainage activity are located in the east (Robinson and Armstrong, 1985).

The soils of the west of the country are moderately well suited to the production of grass. Where the soils are naturally well drained, the combination of climate and soil leads to high levels of potential yields. However, poor drainage status can lead to problems of access and utility. Frequently, then, gley soils in moister areas have high potential yields, but poor trafficability and high poaching risk (Harrod, 1979). It is these soils that might potentially benefit from drainage, but in fact only a small proportion is drained (Forbes et al., 1980).

The major reason for the lack of drainage uptake in grassland areas lies in the economics of grassland production and utilization. As a gross generalisation, the majority of grassland farms are small enterprises with low gross margins and little spare capital for investment and, consequently, there is no competition for the available funds. At the same time, the yield of grass is not normally a limiting factor in the economy of most grassland farms. Imposition of milk quotas have added a further constraint, in that improved profitability cannot be achieved by increased production.

In the light of these problems, it becomes clear that improving the profitability of grassland enterprises requires not an increase in production, but a more cost effective way of achieving similar levels of produktion. Substitution of grazing for more expensive feed-stuffs is one way of reducing costs, and in this context access to the land becomes crucial. Drainage of grassland offers a direct way of improving the length of the access period (grazing season). However, such a drainage benefit cannot be directly quantified (in the way that for example an increase in wheat yield can) and consequently cannot be translated directly into cash terms for investment appraisal without first considering the management system of the farm. Because the benefits of drainage are thus often indirect, and not directly visible in cash terms, the apparent case for investment in drainage is not perceived to be as strong as other claims on limited farm finance.

2. PROBLEMS

Technically, there are few major problems with the drainage of grassland. The major limitation for the adoption of drainage has already been identified as economic. As drainage costs currently average £1000 per ha, such expenditure requires careful investment appraisal. It is the aim of much of the research reported here to provide the tools to permit such an evaluation, through the documentation of the anticipated benefits. Nevertheless, some problems remain, which, while not of major importance, still require solutions. These include:

- The development and validation of appropriate drainage design criteria for grassland. Current recommendations (MAFF, 1982b) are based on a rainfall rate with a one-year return period, a lower rate of acceptance of water by the soil than a comparable site in arable cultivation, and the conventional target of controlling the water table to a level 50 cm below the ground surface. How far these criteria are adequate and appropriate is currently being tested.
- Much grassland is found in upland areas which rarely experience a significant summer soil moisture deficit (e.g. Thomasson, 1981), and many have shallow stony soils which are inherently unstable. Where such conditions occur in areas of low economic return, there is a need for low cost drainage systems. However, the obvious low-cost drainage technique, mole drainage, is unlikely to be effective in these situations because of the nature

of the soil and the difficulty of finding a suitable time to install the work. Low cost drainage for use in upland pastures needs further investigation (Harris et al., 1984a).

- Much grassland would appear to respond to subsoiling treatments, but farmers are reluctant to carry out this treatment because of the uneven surface it creates. Such unevenness may remain for considerable periods in the absence of further cultivations, and hinder further access to the land and, in particular, makes the conservation of grass (taking silage and hay making) difficult. There is a need for further research into effective subsoiling implements that leave a flat surface behind.

3. EFFECTS

In view of the slow uptake of grassland drainage, the Field Drainage Experimental Unit (FDEU) of the UK Ministry of Agriculture, Fisheries and Food (MAFF) is engaged in a research programme with the dual aim of identifying the actual benefits of drainage and their potential value to the farm enterprise, and of making positive recommendations for the improvement of drainage practices. In addition, the information gathered from such experiments is becoming increasingly valuable in the context of the conservation of wet areas, where quantification of the benefits foregone may be required for compensatory payments, or where positive water management is necessary.

Drainage can only remove water under gravity. Its main effects are thus to be seen in the soil water regime of the site, that is the length and duration of saturation (waterlogging) of the soil. The main tool for the study of drainage effects is thus the examination of the soil water regime of experimental sites. Two levels of investigation have been used: firstly the experimental site, where replicated plots are instrumented to provide detailed measurements of soil water status and drain flow, the effects in terms of grass yield or animal production. Secondly the survey site, where a lower level of instrumentation and experimental effort is repeated at multiple locations. Such survey sites provide one means of extending the detailed results to a wider range of conditions.

3.1. Hydrologic effects

Drainage of a soil is basically a hydrologic treatment. It removes excess water to reduce the period of waterlogging. Study of drainage effects must thus take as its starting point the alteration of site hydrology. Over

TABLE 1   Water tables, runoff and nitrate leaching losses, Cockle Park

Drainage treatment		Arable			Grass		
		N	Q	W	N	Q	W
Mole drainage	- drain flow	17.4	233	50	0.2	232	33
	- surface flow	1.4	47		0.0	35	
Control	- surface flow	17.7	224	25	0.6	308	14

Notes: N = nitrate leaching, kg/ha of N
     Q = total runoff, mm
     W = mean water table level, cm below ground level
  arable = fallow following harvest of spring barley (Jan. 1980-Mar. 1980).
  grass = first year grass ley (Nov. 1980-April 1981)

the past few years, three major grassland sites have been the subject of detailed investigation: Cockle Park, Hayes Oak, and North Wyke. Complementary sites in arable cultivation have also been investigated by FDEU (e.g. Harris et al., 1984b).

The Cockle Park experiment was set up in 1978 on the University of Newcastle's experimental farm, and after a period of arable use, is now in a short term grass ley. Details of the site and the experimental work are provided by Armstrong (1984). Over a five year period, it has been consistently observed that mole drainage has significantly lowered the mean water table levels compared to the undrained control (Table 1). Discharge of water from the surface of the undrained area was however equal to the drain flow, so that while drainage had a major effect on the internal hydrology (i.e. soil water regime) of the site, it had little effect on its external behaviour (i.e. generation of runoff). The effect of drainage on the leaching of nitrate N from this site was also studied during the transition from arable to grass, and it has shown firstly that surface and plough layer flows carry as much total nitrate N as the drain flows and, secondly, that the loss of nitrate N from grass receiving low levels of applied nitrogen (75 kg N/ha) is very small in the following winter (Table 1, Armstrong et al., 1983). The data from this site have also shown that on this soil, a clay loam of the Dunkeswick series, the traditional drainage practice of drains alone at a spacing of 12 m has little effect on the soil water regime but that mole drains achieve the desired control of the water table. The future recommendation is that soils of this type should be mole drained.

This site has also been used for the development of techniques for the recording of poaching damage (Section 3.2).

The Hayes Oak experiment provides a comparable experiment on a Denchworth clay soil in the English midlands. Detailed studies of the hydrology, including soil water regime and drain flow, have been reported by Arrowsmith (1983). This site has also provided information on grass yields (discussed in Section 4.2). As at Cockle Park, the effect of mole drains has been to control the water table to depths of 40-50 cm below ground surface, while the mean water table in the control plot has been close to the surface. Peak runoff rates from the undrained plot have been as high as 4 mm/hr, even on quite low gradients (1.5%), and in excess of the rates observed from the drained plots. Total flow volumes, however, are very similar to those of the drained plots (e.g. Arrowsmith, 1983, Figure 7). Such a 'flashy' response can be attributed to the saturated state of the undrained plot, which has a water table very close to the surface, and can offer no storage for incident rainfall.

A major experiment to evaluate both the hydrologic and grass production aspects of drainage was commenced in 1982 as a joint venture between FDEU and the Animal and Grassland Research Institute (AGRI) at North Wyke, Devon. This is an attempt at a comprehensive examination of all aspects of grassland utilization on replicated drained and undrained areas to include: grass production, animal production, sward composition changes, nitrogen cycling, as well as the hydrologic studies of soil water regime and runoff generation. Further details of the experiment, the site, the drainage installation, instrumentation, and preliminary results are given by Armstrong et al. (1984, 1985). Figure 1 illustrates the hydrologic data that are available from this and other sites. In it, hourly rainfall rate, water table levels and the discharge of water is reported for a 7 day period, 20-27 November 1984, for adjacent drained and undrained plots. The surface runoff from the undrained plot is equal to the drain flow from the drained, whereas the drained plot shows the ability of the drainage system to control the water table to a depth of 50 cm. This combination of rapid runoff associated with rapid changes of water table level implies a soil physical system in which water moves rapidly, but with a very low drainable porosity. This sort of response can then be seen as direct evidence of the importance of macropores for soil water movement.

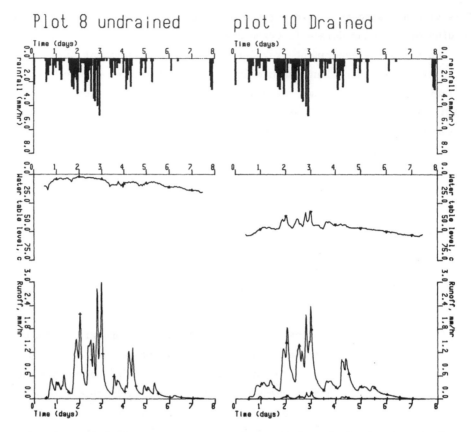

Fig. 1   Rainfall, water table, and runoff from drained and undrained plots, North Wyke, Devon, 20-27 November 1984

3.2. Grassland utilisation

Crucial to the utilisation of grassland is the bearing capacity of the soil. If animals graze a weak soil they deform the surface and damage the sward. This damage is referred to as 'poaching' (Patto et al., 1978). The degree of poaching damage is largely dependent upon soil water content, which in turn is related to the depth to the water table. Laboratory work by Davies (1983) has shown that tensions of 50 cm were sufficient to reduce sinkage of an applied soil load. Length of time at tension has also an important effect of soil strength (Davies, 1985a), showing the importance of drainage in the rapid removal of excess water. This correlates with other studies which have suggested that controlling the water table to 50 cm below the surface is sufficient to reduce poaching.

More difficult is the operational definition of poaching. Subjective assessments in terms of sward damage are common, but objective measures are required. Observing that damage is associated with the indentation of the soil by the hoof of the grazing animal, it has been postulated that recording the surface roughness of the soil can be used to characterise the surface, and hence lead to an objective measure of poaching. Use of a profile meter to record the microtopography of the soil surface over 1 m long transects has been used on the Cockle Park drainage experiment by Armstrong and Davies (1985). Statistical processing of the results is necessary, and it was found that the standard deviation of the profiles after the removal of the overall trend by a third degree polynomial gave a measure that correlated well with the subjective assessments. The results from Cockle Park showed that poaching profiles had standard deviations in excess of 1 cm, sometimes in excess of 2 cm, whereas unpoached sites had values below 0.7 cm (Davies, 1985b).

4. BENEFITS

4.1. Detailed studies

All the sites discussed above have clearly shown that drainage can lead to changes in the soil water regime. Equally, it has been shown that such changes can improve grassland productivity, both in terms of utilisation and herbage production and quality (Jackson, 1982).

Yield of grass was recorded from the Hayes Oak experiment, as were sward composition and nitrogen uptake (Parker, 1983). Although the overall yield increase attributable to drainage was modest (15% in dry matter production), this was concentrated in the early part of the grazing season, when much larger yield increases (of the order of 30%) were recorded. The drainage benefit is thus most marked at a time that is critical in the farming calendar. The sward on the drained land was observed to have a lower proportion of weed species, and a more efficient use of applied nitrogen fertilizer. A simple economic analysis based on the silage value of the increased production shows an estimated benefit of £120 per hectare per annum, compared to an annual cost of £23 per hectare per annum.

Such analyses do not, however, consider the value of the increased grazing period.

4.2. Survey sites

Information about the drainage benefits obtainable from a large number of sites has been obtained from a study of soil water regimes (Armstrong,

1985; Robson and Rands, 1985). For this study a large number of paired drained and undrained areas were studied. Weekly observation of water tables in open auger holes and a walk-over assessment of soil conditions were recorded over a period from 1974-1979. Unfortunately the data quality is not high, as sites were selected on an opportunist basis. In particular problems were encountered with determining the status of the drained plots (and in particular that of any secondary drainage treatments), and also in establishing the comparability of the undrained control. Frequently the response of the farmer to the drainage of the land was to change the land use (e.g. by changing to arable cropping) or to undertake drainage of the undrained area once the drainage effect was demonstrated. Nevertheless, a reliable body of data was obtained from the initial data set, and the results discussed here were derived from a subset restricted to gley soils under grass, a total of 116 site years. Drainage benefits were recorded using as an index the number of days the water table was within 40 cm of the surface (the W40 days index used by Soil Survey of England and Wales for the classification of soil wetness) (Robson and Thomasson, 1977), and also by the number of days the walk-over assessment recorded the site as capable of supporting grazing (grazing days). An overall correlation between soil water regime and grazing period was observed, and in general, the transition from grazing to ungrazable corresponded to a water table at around 50 cm below ground level. This is excellent confirmation of the suitability of the design criteria.

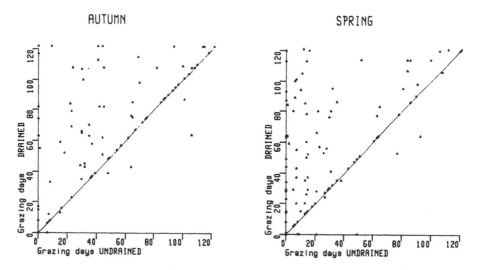

Fig. 2   Observed grazing days on paired drained and undrained plots, for autumn (September-December) and spring (January-April)

Benefits observed varied enormously. Figure 2 shows a scatter diagram
plotting the number of grazing days on the undrained plot against that on
the drained. Data plotting to the left of the line of equality show a drainable benefit, those to the right, a loss. Data are plotted for the autumn
(September-December) and spring (January-April) periods separately. On some
sites the undrained land appeared to perform better than the drained, and on
others the two behaved similarly. This reflects a combination of poor artificial drainage, management problems, and poor site identification. Equally,
some sites showed enormous benefits, changing a situation of no workable
days to continuously workable. Mean values for the increase in work days
were 19 in the autumn and 28 in the spring. Despite all the problems with
the data set, it is clear that drainage benefits can be large, and the utility of the extra grazing, particularly in spring which corresponds to the
period of increased grass growth observed on drained sites.

5. CONCLUSION

The data reviewed here show that drainage can have significant benefits
for grassland production. Not only does it affect the soil water regime by
reducing the duration of waterlogging, but it also leads to an increase in
the grass production, the length of the grazing season, the quality of the
sward, and a reduced poaching risk. Although continued work is necessary to
quantify these benefits, they do offer the farmer a tool for increasing the
productivity of wet land. The lack of uptake of drainage in grassland is
thus due to the twin problems of the availability of capital for investment,
and the perception of the drainage need within the framework of the individual farm enterprise

ACKNOWLEDGEMENTS

Thanks are due to all my colleagues at FDEU for making their data available for this paper, particularly to Pat Davies, and to D.A. Castle for his
constructive comments on the manuscript.

REFERENCES

Armstrong, A.C. 1981. Drainage statistics 1978-80. FDEU Technical Report
      80/1, MAFF, London.
Armstrong, A.C. 1984. The hydrology and water quality of a drained clay
      catchment, Cockle Park, Northumberland. In 'Catchment experiments in
      Fluvial Geomorphology' (eds. T.P. Burt and D.E. Walling), Geobooks,
      Norwich.

Armstrong, A.C. 1985. The fractal dimension of some transient soil properties (in preparation).

Armstrong, A.C., Shaw, K. and Wilcockson, S.J. 1983. Field drainage and
nitrogen leaching: some experimental results. J. Agric. Sci., Camb.
101: 253-255.

Armstrong, A.C., Atkinson, J.L. and Garwood, E.A. 1984. Grassland drainage
economics experiment, North Wyke, Devon. First report: January-Septem-
ber 1982. LAWS Report RD/FE/23, MAFF, London.

Armstrong, A.C. and Davies, P.A. 1985. Surface rogosity as an index of
poaching (in preparation).

Armstrong, A.C., Garwood, E.A., Hallard, M., Talman, A. and Tyson, K. 1985.
Grassland drainage economics experiment, North Wyke, Devon. Report for
1982-4 (in preparation).

Armstrong, A.C., Robson, D. and Rands, J. 1985. The national soil water
regime study, DW1. (in preparation).

Arrowsmith, R. 1983. The hydrology of the Hayes Oak Farm experimental site
1979/82. LAWS Report RD/FE/16, MAFF, London.

Davies, P.A. 1983. The effects of drainage in alleviating grassland poaching. LAWS Report RD/FE/15, MAFF, London.

Davies, P.A. 1985a. Influence of organic matter, moisture content and time
after reworking on soil shear strength. J. Soil Sci. (in press).

Davies, P.A. 1985b. Field measurements of grassland poaching (in preparation).

Forbes, T.J., Dibb, C., Green, J.O., Hopkins, A. and Peel, S. 1980. Factors
affecting the productivity of permanent grassland. Permanent Pasture
Group, Hurley.

Harris, G.L., Morgan, P. and Jones, A.O. 1984a. The modification of surface
layer flows on sloping land by shallow drainage treatments. LAWS Report
RD/FE/20, MAFF, London.

Harris, G.L., Goss, M.J., Dowdell, R.J., Howse, K.R. and Morgan, P. 1984.
A study of mole drainage with simplified cultivation for autumn-sown
crops on a clay soil. 2. Soil water regimes, water balances, and nutrient loss in drain water, 1978-80. J. Agric. Sci., Camb. 102: 561-
581.

Harrod, T.R. 1971. Soil suitability for grassland. In 'Soil Survey Application' (eds. M.G. Jarvis and D. Mackney). Soil Survey Technical Monograph 13, Harpenden.

Jackson, M.V. 1983. Drainage of grassland - note on ADAS experimental work
in South West region. LAWS Report TN/FE/04, MAFF, London.

MAFF. 1982a. Agricultural Statistics, United Kingdom, 1980 and 1981. HMSO,
London.

MAFF. 1982b. The design of field drainage pipe systems. Ministry of Agriculture, Fisheries and Food, Reference Book 345. HMSO, London.

Parker, A. 1983. Benefits and economics of grassland drainage. Interim report for August 1979-November 1982. LAWS Report RD/FE/14, MAFF, London.

Patto, P.M., Clement, C.R. and Forbes, T.J. 1978. Grassland poaching in
England and Wales. Permanent Pasture Group, Hurley.

Robson, J.D. and Thomasson, A.J. 1977. Soil water regimes. Soil Survey
Technical Monograph 11, Harpenden.

Robinson, M. and Armstrong, A.C. 1985. Maps of underdrainage 1971-80 (in
preparation).

Thomasson, A.J. 1981. The distribution and properties of British soils in
relation to land drainage. In 'Land Drainage' (ed. M.J. Gardiner),
Balkema, Rotterdam.

Effects of saturation on pasture production on a clay loam pseudogley soil

J.MULQUEEN
Agricultural Institute, Ballinrobe, Co. Mayo, Ireland

ABSTRACT

An experiment on the effects of continuous saturation on pasture production is described. Saturation reduced annual yields by from 3 to 31%. It also resulted in a shift of peak growth from spring - early summer to late summer - early autumn. Saturation adversely affected perennial ryegrass and encouraged rushes. Nitrogen in part substituted for drainage but its efficiency was reduced by saturation.

1. INTRODUCTION

About 29% of Ireland consists of wet mineral lowland soils and there are also large tracts of peaty and mineral lowlands associated with the 30% of mountain, hill and bogs (Gardiner and Ryan, 1969). By far the greatest proportion of this wet land is a pseudogley with a perched water table overlying a tight gravelly, sandy, silty or clay loam or clay. Commonly a 15 cm topsoil with an organic matter content of 10 to 20% rests on a 10 cm weakly structured intermediate layer which in turn rests on a massive and tight impervious subsoil. The hydraulic conductivity of subsoil is less than 1 mm per day while that of the topsoil is about 0.30 m per day (Burke et al., 1974). In high rainfall areas the topsoil may be more peaty and some soils are now cut over peats. Soils with impervious layers at shallow depth have very limited water storage and they become already saturated and waterlogged by small rainfalls.

The Irish climate is characterised by high rainfalls and low potential evapotranspiration by West-European standards (Figure 1). When the effective rainfall that is the rainfall less the estimated surface run-off is considered, the surplus of effective rainfall over evapotranspiration in the winter season at Ballinamore, Co. Leitrim is about twice the actual balance for central Netherlands. Moreover, potential evapotranspiration exceeds rainfall for three times as long in the Netherlands as in Ireland.

The combination of shallow impervious layer and wet climate results in prolonged saturation of the soil. When the soil is saturated, pasture growth is reduced and there is invasion of water tolerant weeds. There is also reduced utilisation of the pasture caused by too low a bearing capacity and

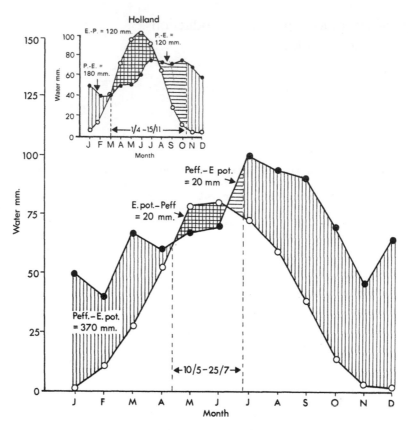

Fig. 1 Effective rainfall (total rainfall less estimated surface run-off) and potential evapotranspiration on a pseudogley at Ballinamore (County Leitrim) compared with total rainfall and potential evapotranspiration in central Holland

shear strength. Cattle and cows walk the pasture into the soft soil which deforms under constant volume and machinery becomes immobilised due to sinkage and wheel spin. Mole or gravel mole drains are the standard drainage solutions and contour drains with loosening are also used in some sloping lands.

This paper describes an experiment to determine the effects of water-logging, drainage and nitrogen on total and seasonal pasture production on a pseudogley in north-central Ireland.

2. LITERATURE REVIEW

Van 't Woudt and Hagan (1957) and Wesseling and Van Wijk (1957) reviewed crop responses at high moisture levels and the effect of depth to water table on crop growth respectively. Williamson and Kriz (1970) reviewed crop

response to flooding and depth of water table. Later Wesseling (1974) reviewed crop growth and wet soils. These reviews show that the effects of high
water tables on crop growth depend on the soil, the crop, the period of the
year in relation to crop growth, the rainfall especially whether the growing
season is wet or dry, the duration of waterlogging, the ambient air and soil
temperature and the level of nitrogen fertilising.

Clay and heavy loam soils require deeper water tables for optimum crop
growth than sandy soils. Shallow rooted crops such as grass and clover are
most tolerant of high water tables and yield best with water tables about 30
cm deep. Damage by saturation depends on both the period of the year in relation to crop growth and the crop. Grasses are more tolerant of high water
tables during the growing season than most other crops. Deeper water tables
appear to be best for most crops in wet growing seasons. Wet soils have high
heat capacity and they heat up slowly. Growth and all temperature dependent
reactions are slowed down. Saturated soils are oxygen deficient. This results in impaired mineralisation of organic nitrogen and some nitrogen is
lost through denitrification.

3. EXPERIMENTAL

The design was a randomised block with 4 treatments viz. 0, 0.37, 0.75
and 1.15 m design depth of water table. Each treatment was replicated 3
times. Plots were on flat ground and were 2.62 m square and each was sur-
rounded with a gravel filled ditch in which a constant water level was maintained by a ball-cock and overflow arrangement. A mole drain was drawn down
the centre of each plot at a depth of 45 cm in dry conditions. Plots were
overseeded with perennial ryegrass and white clover after precutting to overcome some slight damage caused during the preparation. The following fer-
tilisers were applied:

1967 - 6.7 t/ha ground limestone
1967
1968
1969   36 kg/ha P + 110 kg/ha K annually
1970
1967
1968   52 kg/ha N at the beginning of each growth period
1969   52 kg/ha N in spring and 26 kg/ha N at the beginning of each subse-
1970   quent growth period

In 1971 each plot was split in random fashion and 3 levels of N applied
viz. 0, 43 and 86 kg/ha at the beginning of each of 6 growth periods. A 1.82

m wide strip of each plot was harvested, weighed and sampled. Plots were harvested 4 times in 1968, 3 times in 1969, 4 times in 1970 and 6 times or at approximately monthly intervals in 1971. At the same time another experiment on the hydrology of this pseudogley was run closeby and results are given by Burke et al. (1974).

4. RESULTS

4.1. Annual production

Table 1 shows annual productions over the four complete years 1968-1971. Only in 1971 were the drainage treatments significant at the 5% level and only the water table at ground level differed significantly from the deeper water tables. The trend in the other years was toward higher production by keeping the water table at 37 cm or deeper. The average increase in production of the drained treatments over the saturated treatment was 10.5% in 1968, 12.9% in 1969, 2.5% in 1970 and 31% in 1971. In 1971 all plots were sprayed with weed killer to eliminate rushes which were more frequent in the saturated treatments. This is partly responsible for the higher response to drainage in 1971.

TABLE 1   Annual herbage production on 4 design water tables, kg/ha dry matter

Year	Water level				Significance
	0	37	75	115	
1968	8,578	8,109	9,667	10,666	
1969	8,245	8,581	9,287	10,058	
1970	10,126	10,654	9,937	10,543	
1971	8,812	11,117	11,536	12,088	*

4.2. Seasonal production

In all years there was a big response to lowering the water table to 37 cm or greater in the early harvest (Table 2). This response was significant in the May, June and July harvests in 1971 and almost significant in the other years. In the rest of the year there was no significant response but the trend was for an increase in production with lowering the water table to 37 cm or greater. In two harvests after a wet growing period there was a significant response to lowering the water table e.g. the harvest of 30 August 1968 when the average response to lowering the water table from the surface was 23% from 2720 to an average of 3336 kg/ha.

88

TABLE 2   Production of dry matter in the 1st harvest (kg/ha)

Date	Water level				F*
	0	37	75	115	
24-5-68	2289	2344	2784	3182	2.18
4-7-69	4013	4226	4786	5551	2.56
13-5-70	1511	1696	1575	1727	3.23
5-5-71	910	2108	2161	2175	10.2

*F value at 5% point = 4.76

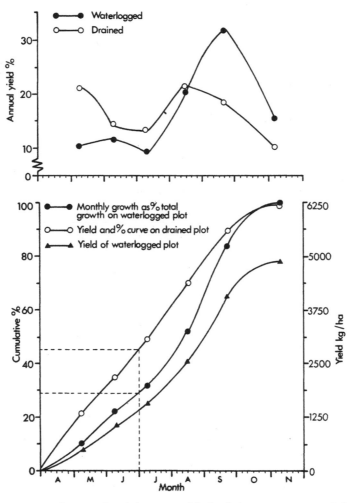

Fig. 2   Seasonal growth with no applied nitrogen expressed in kg/ha and as percentage of total for year on the saturated treatment compared with the 37 cm water level treatment

When the number of harvests were increased to six in 1971 it became possible to compare the seasonality of growth curves on saturated plots and drained plots. This is shown in Figure 2, which shows that apart from the 24% lower production, the saturation resulted in a shift in the seasonality of production. On 1 July only 28% of annual production had grown on the wet treatment compared to 46% on the drained treatments. Saturation delays the peak growing season and this was only partly offset at very high N-levels.

4.3. Effect of nitrogen

A preliminary test along with observations on growth suggested that the nitrogen fertilising was influencing the results and could be masking the effects of saturation. Accordingly the drainage treatments were split and three levels of nitrogen applied. These showed that the nitrogen gave very highly significant responses at both levels over the control in all harvests. Only in the last two harvests there was no response to the higher N level and there was a significant depression over the lower level in the last harvest. The results are shown in Table 3.

TABLE 3   Response to three levels of N

Harvest	N level (kg/ha)			Significance
	0	43	86	
5- 5-71	1183	1,920	2,413	***
8- 6-71	965	1,880	2,543	***
8- 7-71	780	1,859	2,567	***
15- 8-71	1228	2,606	3,085	***
21- 9-71	1359	2,508	2,565	***
5-11-71	790	1,359	1,057	***
Total	6304	12,132	14,230	***

As already noted the drained treatments were significantly better than the saturated treatment in the May, June and July harvests and this was also reflected in the annual yield. The water level by nitrogen interaction was only significant in the June harvest and was close to significance in the 1st, 5th and 6th harvests. The total production is shown in Table 4. It is apparent that nitrogen can substitute for drainage and that even at 258 kg/ha N it can substantially outyield the best drainage treatment at zero N input.

TABLE 4   Dry matter production under 4 water levels and 3 N-levels (kg/ha)

Water level*	N***		
	0	43	86
0	4869	9,614	11,955
37	6247	12,212	14,892
75	6633	12,724	15,253
115	7469	13,976	14,821

*significant at 5%; ***significant at 0.1%

TABLE 5   Increase in production from applying 43 kg/ha nitrogen at the beginning of each growth period (kg dry matter/kg N)

	Yield				
WT	5.5	8.6	8.7	15.8	total
0	6.8	9.9	20.3	36.4	18.4
0.37	20.6	20.8	19.0	22.9	23.1

While nitrogen can overcome the effects of saturation to a large extent, its efficiency is lower under saturated conditions (Table 5). The efficiency of applied nitrogen on the saturated treatment was less than 50% of that on drained treatments for the first two harvests. In the 3rd and 4th harvests the saturated treatment caught up and exceeded the drained treatment in the 4th harvest. Overall, there was a drop of 4.7 kg dry matter/kg N in the efficiency of nitrogen on the saturated treatment.

4.4. Seasonal response to water table at the 3 levels of N

Taking the yield at the 37 cm water level at 100 Table 6 shows that the effect of saturation varied. In the early part of the growing season the effect of saturation was to depress growth though less so at the high level (86 kg/ha N at each application). From mid-August onwards saturation resulted in increased growth on the zero nitrogen treatment. Overall the production on the saturated treatment was 80% that on the 37 cm deep water level treatment.

4.5. Botanical effects

Samples were separated into perennial ryegrass, rushes (Juncus species) and other herbs in 1968 and 1969 after one and two full years of operation.

TABLE 6   Seasonal response to saturation taking the yield at the 0.37 m deep level at 100

Harvest		N	
	0	43	86
5- 5-71	39	36	51
8- 6-71	63	55	63
8- 7-71	55	82	89
15- 8-71	75	113	103
21- 9-71	132	94	86
5-11-71	116	93	103
Total	81	79	80

Saturation significantly reduced the percentage of perennial ryegrass in both years from 60% down to 37% in 1968 and from 86% down to 58% in 1969. Rushes were significantly higher in the saturated treatment, 26% compared with 7% and 18% compared 6% in 1968 and 1969 respectively. Saturation reduced the useful grasses and increased the rush content.

5. DISCUSSION AND CONCLUSIONS

The overall response of pasture to lowering the water level from ground level to 37 cm or more was relatively small varying from 2.5% to 31%. In the adjacent hydrology trial of Burke et al. (1974) where mole drainage was compared with no drainage on a slope, the response to mole drainage was 9%, 14% and -3% in 1967, 1968 and 1969. This indicates that pasture species are fairly tolerant of prolonged saturation. Botanical measurements, however, indicate that perennial ryegrass was adversely affected by the saturation while rushes were favoured. This is in spite of the well-known fact that rushes are severely checked and controlled by frequent cutting.

Another very significant result was the effect of saturation on seasonal production. Saturation resulted not only in reducing overall growth but it delayed peak growth and greatly reduced early growth. Very significantly also it reduced the response to applied nitrogen in the first half of the growth season although nitrogen did substitute for drainage also. In the hydrology trial the responses to mole drainage were 34% and 15% for harvests cut before mid-May. The result on seasonal production is of significance for farming for which optimum growth conditions in the April to June period, the period of maximum pasture growth and conservation, is most important.

Water levels below 37 cm were not significantly better than the 37 cm depth. However, because of the very low hydraulic conductivity of this soil whose pores at depth are practically all capillary size, it is unlikely that the water table was at the water level in the external drain. It appears that the treatments with external water levels of 75 and 115 cm would correspond to mole drained farm land while the treatment with a water table held at 37 cm would have a slightly higher water table than mole drained farm land in rainy weather and much higher in dry weather. The saturated treatment is representative of a very wet growing season.

In measuring response to drainage by pasture, it is important to measure production frequently at critical times of the year and this is the first half of the growing season when growth is at a peak. Since nitrogen can partly compensate for drainage and more so at high application rates, it is also important to include nitrogen treatments.

Apart from the growth of grass, its utilisation is at least as important. It is well-known that animals and machinery need top soil conditions with a high bearing capacity and shear strength. Grass utilisation on the saturated treatment by livestock would be low as it would be trampled into the soft ground, ground cover would be reduced and growth conditions would deteriorate. Machinery would be liable to sink and low shear strength would result in wheel spin and consequent soil damage. Saturated soil conditions have the 3-fold effects of reduced annual growth, a seasonal shift in peak growth to late summer and autumn and greatly reduced utilisation of the herbage.

REFERENCES

Burke, W., Mulqueen, J. and Butler, P. 1974. Aspects of the hydrology of a gley on a drumlin. Ir. J. Agric. Res. 13: 215-228.
Gardiner, M.J. and Ryan, P. 1969. A new generalised soil map of Ireland and its land-use interpretation. Ir. J. Agric. Res. 8: 95-109.
Van 't Woudt, B.D. and Hagan, R.M. 1957. Crop responses at excessively high soil moisture levels. In 'Drainage of agricultural lands' (ed. J.N. Luthin). Agronomy 7: 514-578, Amer. Soc. Agron., Madison, Wisconsin.
Wesseling, J. and Van Wijk, W.R. 1957. Soil physical conditions in relation to drain depth. In 'Drainage of agricultural lands' (ed. J.N. Luthin). Agronomy 7: 461-504, Amer. Soc. Agron., Madison, Wisconsin.
Wesseling, J. 1974. Crop growth and wet soils. In 'Drainage for agriculture' (ed. J. van Schilfgaarde). Agronomy 17: 7-37, Amer. Soc. Agron., Madison, Wisconsin.
Williamson, R.E. and Kriz, G.J. 1970. Response of agricultural crops to flooding, depth of water table, and soil gaseous composition. Amer. Soc. Agr. Eng. Trans. 13: 216-220.

Discussion Session 2.2

Question from Mr. Ph. Lagacherie to Mr. A. Armstrong

In the Cockle Park experiment, you have compared mole drainage with 12 m spacing drain. I would like to know whether you have observed significant differences in poaching or crop yields.

Answer of Mr. Armstrong

We have no crop yield data from this experiment. Observations of poaching damage have shown that the drains-only system occupies a position intermediate between the mole drainage system and the control. Although the drains-only system has only small effects on the water table, it does appear to have the effect of reducing the build-up of surface water at the lower end of the slope. The surface water appears to be critical to the incidence of poaching damage, and thus by reducing this component of site hydrology, drains alone have an effect on the grass utilization which is greater than a simple examination of water tables might suggest. However, this observation does not invalidate the conclusion that mole drainage is the most effective technique for the control of both water tables and poaching damage on this site.

Question from Mr. B. Lesaffre to Mr. A. Armstrong

In your paper, you give information on the comparison between drained and undrained areas. The drained land does not perform better than the undrained, on the same site. Moreover, there are big differences between the drained sites. What is in your opinion the most important one of the three reasons you mention (poor artificial drainage, management problems, poor site identification)?

Answer of Mr. Armstrong

Some of the variability within the data set is due to poor identification of the drained/undrained pairs that make up the observation points. It is suspected that the majority of the cases where the undrained site appeared to perform better than the drained, is due to soil differences between the two-component observation points. However, we do have one data point where poor performance of the drained site has been documented as being a consequence of injudicious site management. Nevertheless, the majority of data points which show only small drainage benefit seem arise from ineffective drainage, and we must therefore conclude that many of the drainage schemes surveyed do not give the expected

95

benefit. This may be due to the use of inappropriate drainage techniques, and in particular the failure to renew secondary treatments (mole drainage or subsoiling). Lastly, the data refer to the period 1974-1979, and include a proportion of results from the extremely dry winter of 1975/76 where on several sites no drainage effect was observed because no drainage stress experienced.

Question from Mr. P.E. Rijtema to Mr. J. Mulqueen

It is well-known that poor drainage conditions can be overcome by additional N-fertilization. Are the production data you showed gross productions, or did you take into account losses by grazing cattle? In discussions on the effect of improvement of drainage of grassland in our country the aspect of production losses due to poor drainage has become very important. Do you have any data on these production losses in order to translate gross production into net production?

Answer of Mr. Mulqueen

No, we have not measured the utilisation because of the great difficulties involved. We know the effects include walking the grass into the soil, as a result of deformation the soil at the base of the hoof mark, bare reducing of the growing area and the conditions for pasture growth and an increased growth of weeds. Machinery tends to wheelspin or sink and soil is mixed with forage silage causing great damage to silage quality.

Simulating effects of soil type and drainage on arable crop yield

A.L.M.VAN WIJK & R.A.FEDDES
Institute for Land & Water Management Research (ICW), Wageningen, Netherlands

ABSTRACT

To evaluate effects of drainage on crop yield physical and biological processes occurring in spring, growing season and autumn have been studied in detail. This paper deals with the influence of drainage upon:

- groundwater table depth and waterlogging during winter time;
- number and time of occurrence of workable days for sowing and planting operations;
- duration of germination and time of emergence in dependence on soil water water content and -temperature;
- water uptake, development and growth of crops.

A number of relationships and models have been developed and combined into an integrated model approach that considers all those effects. The model operates on a daily basis using actual weather and soil data. It has been tested for various conditions in the field.

Results of 30 years of model simulation on potatoes grown on two soil types for five different drain depths and three drain intensities are presented.

1. INTRODUCTION

In the Netherlands land development projects are executed over an area of about 35 000 ha yearly. For the economical evaluation of these projects quantitative information is needed on the effects of soil and water management measures upon agricultural crop yield. Up till now these effects are determined in a more or less empirical way.

During the past decade more and more effort has been put into the development of computer models that simulate crop growth and production in dependence of soil and climatic conditions. The advantage of simulation models is that effects of changes in soil/water management can be obtained in a short time over long time periods over which climatic records are available.

During a number of years an integrated model approach that quantifies the effects of water management on crop yield has been developed step by step (see Feddes and Van Wijk, 1977; Van Wijk and Feddes, 1982). The present paper describes the most recent state of the art of this approach. The model predicts effects of changes in water management by drainage on trafficability and workability in spring, sowing/planting time, emergence date, transpiration, growth and dry matter yield of crops, for different types of soils under various meteorological years. To illustrate the capability of the model with respect to the quantification of all these aspects, some examples will be given. These examples include the simulation of drainage effects on the production of a potato crop grown during 30 meteorological years on a sandy loam and on a 40 cm loam on sand soil profile.

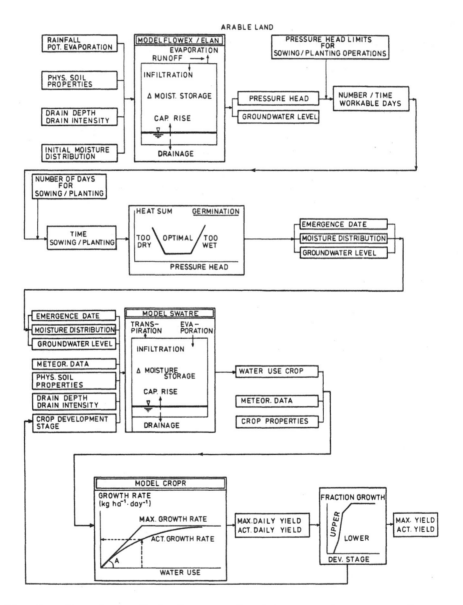

Fig. 1   Flow chart of the integrated model approach for computing the influence of water management on yields of arable land

2. MODELING SOIL WATER MANAGEMENT, CROP GROWTH AND -PRODUCTION

A model that simulates effects of drainage on crop production has to
reckon with effects of both 'too wet' and 'too dry' conditions. Such a
model must be able to indicate quantitatively how soil water conditions in
winter, time of field operations, germination and emergence in spring, crop
growth and production in summer and workability in autumn are influenced by
drainage conditions. In Figure 1 a flow chart of such an integrated model
approach is presented.

2.1. Modeling soil water conditions in winter and spring

The model FLOWEX (Wind and Van Doorne, 1975; Buitendijk, 1984) computes
the terms of the soil water balance of a non-cropped soil profile, that may
consist of different layers. The model has been developed to consider effects of changes in soil water content on the utilization conditions of the
soil, such as workability and trafficability.

FLOWEX is based on an integrated form of the Darcy flow equation assuming an exponential relationship between hydraulic conductivity and soil water pressure head. Combination with the continuity equation results in a
description of the non-steady state flow process, layer by layer for consecutive time increments. As boundary condition at the soil surface rain-
fall/potential soil evaporation is used. As boundary condition at the bottom
a flux - groundwater table depth relationship, based on the drainage theo-
ries of Hooghoudt (1940) and Ernst (1956) is applied.

Input into FLOWEX is: rainfall and potential soil evaporation on a 24-
hour basis, soil water retention and hydraulic conductivity curves for the
different soil layers, drain depth and drain intensity and finally the
initial soil water profile. Output of the model includes all the terms of
the water balance, the distribution of soil water content, pressure head
and fluxes with depth. The suitability of the soil for field operations depends on the pressure head in the top layer of the soil. For this purpose
the pressure head at 5 cm depth is taken.

2.2. Modeling workability in spring

In two rather different meteorological years measurements of the pressure head at 5 cm depth in spring were regularly carried out on a number of
different soils. Simultaneously farmers were asked to give their judgement
of the suitability of the soil for field operations in terms of good, moderate and poor. In Figure 2 the result for a number of soils is shown,

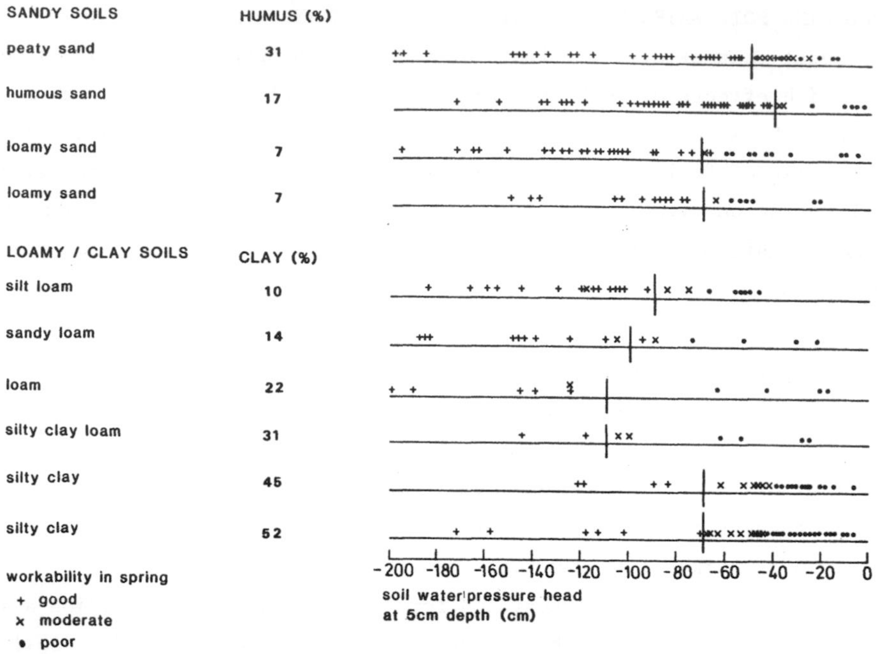

SANDY SOILS   HUMUS (%)

peaty sand   31

humous sand   17

loamy sand   7

loamy sand   7

LOAMY / CLAY SOILS   CLAY (%)

silt loam   10

sandy loam   14

loam   22

silty clay loam   31

silty clay   45

silty clay   52

-200 -180 -160 -140 -120 -100 -80 -60 -40 -20 0

soil water pressure head
at 5cm depth (cm)

workability in spring
+ good
x moderate
• poor

Fig. 2   Farmers' appraisals of workability of a number of different soils for planting potatoes versus the soil water pressure head at 5 cm depth, simultaneously measured in the field

Table 1   Soil water pressure heads at 5 cm depth at which sowing of spring cereals and sugar beets and planting potatoes can be executed without deterioration of soil structure

Soil type	Soil water pressure head (cm)		
	Spring cereals	Sugar beet	Potatoes
Sandy soils	−50	− 70	− 70
Silt/sandy loam (8–20% particles <2 μm)	−80	−100	−100
Loam/silty clay loam (20–40% particles <2 μm)	−60	−100	−120
Silty clay (>40% particles <2 μm)	−40	− 60	− 80

which enables the derivation of pressure head limits that are critical for workability. As a result a summary of pressure head limits for sowing spring cereals, sugar beet and planting potatoes for four main soil groups is presented in Table 1. With the aid of the workability limits in this table, one can derive time and number of days that the soil is workable from a day to day simulated course of the pressure head at 5 cm depth. Knowing the

100

earliest possible sowing/planting date and the number of days required for these operations one can predict the time of sowing/planting.

In the Netherlands the earliest possible sowing date for spring cereals is 1 March. For sowing sugar beets and planting potatoes this date is 20 March. The number of days that are required on an average farm in the Netherlands for sowing and planting is for spring cereals 1, for sugar beets 2 and for potatoes 4 days.

2.3. Modeling germination and emergence

Knowing the sowing/planting date the emergence date can be predicted if one is informed how germination depends on soil water content and temperature in the seed bed. Under optimal conditions of soil moisture time of emergence t can be predicted from the relationship:

$$F = (\overline{T} - T_{min})t \tag{1}$$

where F = heat sum required for a certain percentage of emergence ($^{o}C \cdot d^{-1}$)

$\overline{T}$ = mean soil temperature

T_{min} = minimum temperature (^{o}C) below which no germination occurs

From sowing – emergence experiments in the field one can derive T_{min} and F through eq. (1), by plotting $\overline{T}$ versus $1/t$ (see Feddes, 1971). Often soil temperatures are not available, hence one applies in eq. (1) for $\overline{T}$ the average air temperature. As an example Figure 3 is presented for potatoes. Data in this figure are based on 10 years of field experiments.

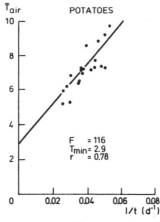

Fig. 3   Mean air temperature $\overline{T}$ plotted versus the reciprocal of time required for emergence of potatoes in order to derive T_{min}. F = heat sum ($^{o}C \cdot days$), r = correlation coefficient

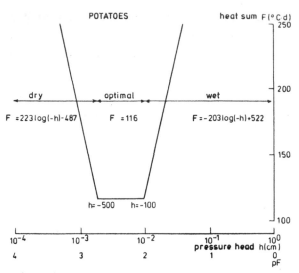

Fig. 4   Relationships between the heat sum required for emergence of potatoes, based on daily mean air temperatures and the soil water pressure head in the seedbed

Under sub-optimal soil water conditions eq. (1) does not apply. Germination is retarded when the seedbed is either 'too dry' or 'too wet'. To model the combined effect of soil temperature and soil water on germination and emergence the effective heat sum can be related to the soil water pressure head of the seedbed. In field experiments Feddes (1971) found for a number of crops a constant sum when the pressure head h ranged between -100 and -500 cm. Outside this range germination was retarded through either abundancy or shortage of soil water. From the same experiments, it appeared that at h = -20 cm and at h = -2000 cm, F $\approx$ 250 $^{\circ}$C·days. The relationships thus obtained and finally used in the model calculations for potatoes are depicted in Figure 4. Similar relationships are available for spring cereals and sugar beet.

Starting at the sowing date a heat sum is now calculated day by day according to the relationship in Figure 4, using the 24-hour mean air temperature and the simulated pressure head at 5 cm depth in the seedbed.

2.4. Modeling crop water use

The model SWATRE (Feddes et al., 1978; Belmans et al., 1983) computes water flow in a heterogeneous soil – root system. Water extraction by roots is accounted for by a sink term. The partial differential equation valid for the flow is approached by a finite difference equation. Taking into account the upper and lower boundary conditions the equation is solved implicitly for each separate soil compartment applying the Thomas algoritme.

Input into SWATRE consists of: initial pressure head distribution with depth, 24-hour data on rainfall, potential soil evaporation and potential transpiration, the soil water characteristics and hydraulic conductivity curves for the different soil layers, rooting depth (varying with time), critical pressure head values for water uptake by roots (sink term) and drain depth and -intensity.

Output of the model includes all the terms of the water balance, the distribution of soil water content, pressure head, fluxes and water uptake by roots in dependency of depth. The main output is the transpiration rate, calculated as the integral of the sink term over the depth.

2.5. Modeling crop development, growth and production

Daily actual dry matter growth rate of a crop having optimal nutrient supply q ($kg \cdot ha^{-1} \cdot d^{-1}$) is calculated with the simulation model CROPR (Feddes et al., 1978) as:

$$(1 - \frac{q}{A \frac{T}{\Delta e}})(1 - \frac{q}{q_{pot}}) = \xi \tag{2}$$

where A is maximum water use efficiency ($kg \cdot ha^{-1} \cdot cm^{-1} \cdot mbar$) determined from field experiments, T is actual transpiration rate ($cm \cdot d^{-1}$) from SWATRE, Δe is vapour pressure deficit of the air (mbar), q_{pot} is potential growth rate ($kg \cdot ha^{-1} \cdot d^{-1}$) computed as a function of among others radiation and leaf area index, ξ is a mathematical parameter ($\xi = 0.01$).

In earlier versions of CROPR, a certain prescribed variation of soil cover with time was assumed. At present the models SWATRE and CROPR are combined into one model, with the development of the crop being generated by the model itself.

The development of a crop with time varies usually from year to year, depending on environmental factors such as temperature, day length, soil water content, etc. To solve this problem time is made dimensionless by introducing the development stage D_s of the crop. If D_s is set 0 at emergence (t_{em}) and 1 at harvest time (t_{ha}) any intermediate development stage is then defined according to:

$$D_s(t) = (t - t_{em})/(t_{ha} - t_{em}) \tag{3}$$

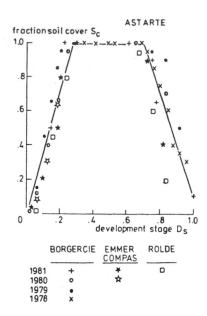

ASTARTE

	BORGERCIE	EMMER COMPAS	ROLDE
1981	+	*	□
1980	o	☆	
1979	•		
1978	x		

Fig. 5   Variation of S_c with D_s for potatoes (variety Astarte) grown at different locations over a number of years

to obtain information about the parameters in eq. (3) for potatoes used here as an example, long year field experiments were analyzed. This analysis showed that variation of soil cover S_c with D_s is constant over the years (Figure 5). Determination of the parameter t_{em} has been described in Section 2.3. In the Netherlands t_{ha} is about 15 September for the variety Bintje and 15 October for the variety Astarte.

Because there is a fixed relationship between leaf area index I (ha·ha^{-1}) and fraction of soil cover S_c (data of Van Loon, pers. comm.[*])

$$I = 2.6S_c + 1.5S_c^2 + 0.9S_c^3 \qquad (4)$$

the variation of I with D_s is known also. Hence q_{pot} can be computed and q be solved from eq. (2).

The increase in total dry matter production as a function of development stage D_s is then distributed over shoot and tubers according to the field experimentally based relationship shown in Figure 6.

Having calculated actual growth rates q^i day by day, final yield Q is

[*]C.D. van Loon. Experimental Station for arable crops and vegetables in the open. Edelhertweg 1, P.O. Box 430, 8200 AK Lelystad

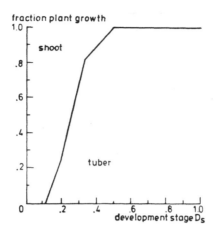

Fig. 6   Distribution of increase in total dry matter production over
shoot and tubers of a potato crop

obtained as:

$$Q = \sum_{i=1}^{n} q^i \Delta t \tag{5}$$

where Δt represents a period of 1 day. In a similar way one can calculate
potential yield Q_{pot}.

3. SIMULATION OF THE EFFECTS OF DRAINAGE ON CROP YIELD

The model has been verified with measurements in the field at different
points in the scheme given in Figure 1. As an example Figure 7 compares the
variation of pressure head at 5 cm depth and the groundwater table depth
simulated with FLOWEX with measured data for the sandy loam soil in the
spring of 1980. An example of the model behaviour during the growing season
is presented in Figure 8. This figure compares the total dry matter and
tuber production of potatoes grown in 1981 on a sandy soil simulated with
SWATRE/CROPR with measured productions. Both figures show that both soil water and plant growth conditions can be simulated reasonably well.

To illustrate the capabilities of the integrated model approach results
of 30 years (1952–1981) of model simulation on potatoes grown on a sandy
loam and a 40 cm loam over sand soil profile are used. The soil water characteristic and hydraulic conductivity curve of both soil types have been
taken from Beuving (1984). Each soil is thought to have a drainage system
installed at depths of 60, 90, 120, 150 and 180 cm. For each drain depth
three different drain spacings, expressed as intensities A (d^{-1}) are applied.

105

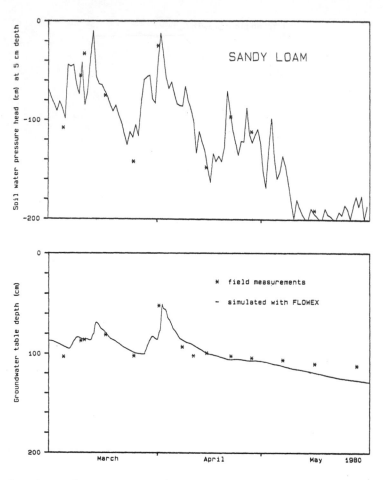

Fig. 7   Comparison of the soil water pressure head at 5 cm depth and
groundwater table depth, simulated with FLOWEX with data measured in the
sandy loam in spring 1980

The value A in fact gives the relation between discharge and height of water
table midway between parallel drains and incorporates the drain spacing L.
According to Hooghoudt (1940):

$$A = \frac{8K_s d}{L^2} \tag{6}$$

where K_s = soil hydraulic conductivity $(cm \cdot d^{-1})$

d = thickness of the so-called equivalent layer (m)

106

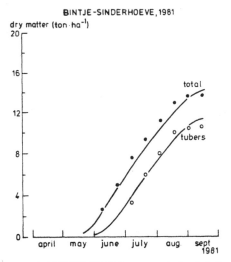

Fig. 8 Comparison of SWATRE/CROPR computed yields (lines) with measured data (points) of potatoes growing in 1981 on a 40 cm humous medium coarse sand overlying very coarse sand

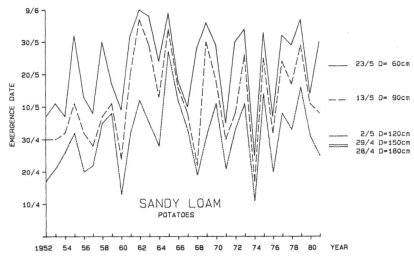

Fig. 9 Yearly emergence dates of potatoes on the sandy loam simulated over a 30-year period at different drain depths

3.1. Influence of drainage on germination and emergence date

In the way described in Sections 2.1 to 2.3 emergence dates were calculated for all combinations mentioned above. Figure 9 shows the emergence date of potatoes grown on the sandy loam soil for each separate year at five different drain depths with comparable drain intensities. It appears to be very effective to increase drain depth. Going from D = 60 cm to D = 90 and

107

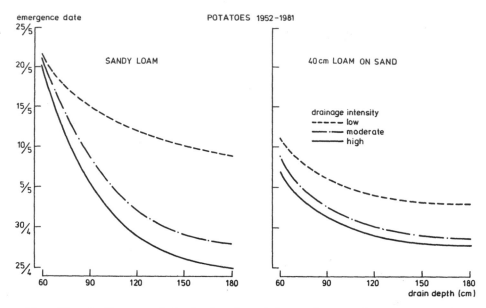

Fig. 10   Influence of drain depth and intensity on the emergence date
of potatoes growing on the sandy loam and 40 cm loam on sand soils,
based on a simulation over the period 1952-1981

D = 120 cm the 30-year averaged emergence date advances with 10 and 21 days.
Increasing drain depth further the advance of emergence date becomes less
progressive. Increase of drain depth goes in general together with deeper
groundwater tables, hence with less potential for capillary rise from the
groundwater and thus more rapid drying of the top soil. Considering the
variations in emergence date from year to year at a certain drain depth,
these can amount up to 1 month. This holds also for the larger drain depths.
Figure 10 summarizes 30-year averages of emergence date of potatoes versus
drain depth at a low, moderate and high drain intensity for sandy loam as
well as 0.40 m loam on sand. For the sandy loam both drain depth and intensity (distance) greatly influence the earliness of emergence. Loam on sand
shows these effects much less pronounced. This difference in behaviour can
be ascribed to the difference in hydraulic conductivity of the sandy loam
soil and the 40 cm loam cover. The lower the hydraulic conductivity of a
soil, the lower the capillary rise from the groundwater table, the less the
soil water conditions at the top are influenced by the groundwater table
depth. In other words: the evaporation rate from the top soil cannot be met
by the water supply from below. The heavier the soil, the more pronounced
this behaviour.

3.2. Influence of drainage on crop yield

Crop water use and dry matter production have been calculated for all chosen combinations of soil, drain depth and -intensity for the 30-year period according to the procedures described in Sections 2.4 and 2.5.

In order to account for the drainage effects that occur separately in spring and in summer the following terminology will be introduced:

Q_{max} – maximum possible dry matter yield that can be obtained under the prevailing weather (i.e. radiation) conditions, assuming no shortage of water for the object with the earliest possible emergence date. For the combinations considered this date is always found at the greatest drain depth, i.e. 180 cm, with the highest drain intensity

Q_{pot} – potential dry matter yield obtained, assuming no shortage of water, starting at the actual emergence date of the considered drain depth/intensity combination

Q_{act} – actual dry matter yield obtained at the prevailing soil water conditions starting at the actual emergence date of the considered drain depth/intensity combination

Expressing separately the drainage effects occurring in spring and summer the following relationship holds:

$$\frac{Q_{act}}{Q_{max}} = \frac{Q_{pot}}{Q_{max}} \cdot \frac{Q_{act}}{Q_{pot}} \tag{7}$$

total=spring. summer

The spring term accounts for the reduction in yield as a result of retardation in the sowing and emergence dates due to too wet soil conditions in spring. The summer term quantifies the shortage of water occurring during the growing period. The total drainage effect including both the influence of earliness in spring and water supply in summer is found by multiplying the spring with the summer-term in eq. (7).

In Table 2 30-years averaged relative yields of potatoes growing on sandy loam and 40 cm loam on sand due to drainage effects in spring, summer and over the total season are presented for five drain depths.

A number of comments can be made:

– the effect of drainage in spring on yield (Q_{pot}/Q_{max}) is much more pronounced on the sandy loam than on the 40 cm loam on sand;

TABLE 2   30-year averages of relative yields of potatoes due to effects of drainage in spring (Q_{pot}/Q_{max}), in summer (Q_{act}/Q_{pot}) and the total combined effect (Q_{act}/Q_{max}) at five different drain depths for a sandy loam and a 40 cm loam on sand

D (cm)	$\dfrac{Q_{pot}}{Q_{max}}$	$\dfrac{Q_{act}}{Q_{pot}}$	$\dfrac{Q_{act}}{Q_{max}}$
Sandy loam			
60	0.83	0.97	0.81
90	0.92	0.95	0.87
120	0.98	0.94	0.92
150	1.00	0.90	0.90
180	1.00	0.87	0.87
40 cm loam on sand			
60	0.93	0.92	0.86
90	0.96	0.90	0.86
120	0.99	0.88	0.87
150	1.00	0.85	0.85
180	1.00	0.81	0.81

- drain depths of 150 cm are minimally required to prevent yield reduction due to retardation of emergence. For explanation of this behaviour see the comments given to Figure 10;
- the reduction in yield due to water shortage during summer (see Q_{act}/Q_{pot}) is largest on the 40 cm loam on sand. The reason for this is the poor capillary supply from the groundwater to the root zone through the sub-soil of sand;
- the reduction in yield due to deeper drainage is in both cases of the same magnitude;
- the optimum drain depth for total yield (see Q_{act}/Q_{max} column) is for both soil types about 120 cm. The relative production level at this depth is however highest on the sandy loam;
- yield reduction on the 40 cm loam on sand is caused mainly by water short-age in summer, while on sandy loam yield reduction is mostly affected by differences in emergence date.

In Figure 11, the 30-year averaged data of Table 2 are presented togeth-er with comparable data for the very wet year 1965 and the very dry year 1976. The 1965-year can be characterized as a 98% dry year, while 1976 is a 1% dry year. For example: a 1% dry year is defined as a year that has an evapotranspiration surplus for the period April – September inclusive that is exceeded only once in 100 years. The characterization holds especially

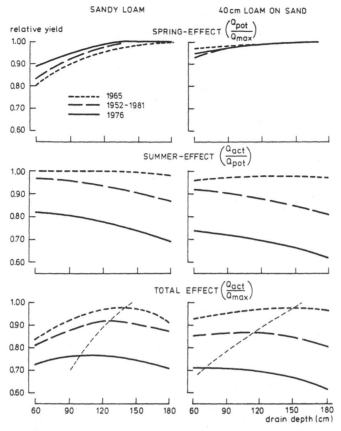

Fig. 11 Relative yield of potatoes growing on a sandy loam and a 40 cm loam on sand profile versus drain depth. Drainage effects occurring in spring, summer and in the total season are presented for the very wet summer of 1965, for the very dry summer of 1976 as well as averaged over the 30-year period 1952-1981

for the summer period and is thus not specific for the spring. A dry/wet spring has no relationship with the dryness/wetness of the summer that follows!

As far as the drainage effect in spring is concerned, the years 1965 and 1976 do not deviate much from the 30-year average. However, other years may give much larger deviations from the average.

The drainage effect on sandy loam in the wet summer of 1965 is negligible, except for a slight reduction at the 180 cm depth. In the dry summer of 1976, however, a yield reduction of 18% occurs at 60 cm depth and of 31% at 180 cm depth. The 40 cm loam on sand shows in 1965 at 0.60 m drain depth a reduction in yield due to poor aeration as a result of too wet conditions. In 1976 the yield reduction at 60 cm drain depth is 26% and at the 180 cm depth 38%.

The part of Figure 11 depicting the total drainage effect on crop yield enables to derive the optimal drain depth in dependence on dryness/wetness of the growing season. For the sandy loam this optimal drain depth varies from 100 cm when based on a very dry year to 140 cm when based on a very wet year. On the average the optimal drain depth is 130 cm. For the 40 cm loam on sand the optima are more widespread. The optimal depths are in dry years 70 cm, in wet years 150 cm and on the average 115 cm.

In this paper the demonstration of the developed methodology has been restricted to the evaluation of drainage effects on the production of a potato crop growing on two soils in the Netherlands. The approach, however, allows for application in other climates for a variety of crops growing on different soil types. Moreover not only drainage effects can be considered, but also effects of irrigation, soil improvement, etc. Therefore the approach is applicable to land evaluation studies in general.

REFERENCES

Belmans, C., Wesseling, J.G. and Feddes, R.A. 1983. Simulation model of the water balance of a cropped soil: SWATRE. J. Hydrol. 63,3/4: 271-286. Techn. Bull. 21 ICW (new series), Wageningen.
Beuving, J. 1984. Soil water characteristics, hydraulic conductivity curves, bulk densities and textures of sand, sandy loam, loam, silty clay and peat soils (in Dutch). Rapport 10 ICW, Wageningen. 26 pp.
Buitendijk, J. 1984. FLOWEX: a numerical model for simulation of vertical flow of moisture in unsaturated layered soils (in Dutch). Nota 1494 ICW, Wageningen. 61 pp.
Feddes, R.A. 1971. Water, heat and crop growth. Thesis, Comm. Agric. Univ. Wageningen 71.12. 184 pp.
Feddes, R.A. and Van Wijk, A.L.M. 1977. An integrated model-approach to the effect of water management on crop yield. Agric. Water Man. 1.1: 3-20. Techn. Bull. 103 ICW, Wageningen.
Feddes, R.A., Kowalik, P.J. and Zaradny, H. 1978. Simulation of water use and crop yield. Simulation monograph. PUDOC, Wageningen. 189 pp.
Wijk, A.L.M. Van and Feddes, R.A. 1982. A model approach to the evaluation of drainage effects. In 'Land Drainage' (ed. M.J. Gardiner). A seminar in the EC Programme of Coordination of Research on Land Use and Rural Resources, Cambridge, UK, 27-31 July 1981. A.A. Balkema, Rotterdam: 131-149.
Wind, G.P. and Van Doorne, W. 1975. A numerical model for the simulation of unsaturated vertical flow of moisture in soils. J. Hydrol. 24: 1-20. Techn. Bull. 101 ICW, Wageningen.

Modelling the yield of winter wheat under limiting water conditions

A.FARSHI, J.FEYEN, C.BELMANS & N.KIHUPI
Katholieke Universiteit Leuven, Belgium

ABSTRACT

The SWATRE and CROPR models are frequently used in combination to define the response of crop growth to water stress. The actual day transpiration, being one of the outputs of the SWATRE model for budgeting the soil water, is used as a primary input of the crop growth model CROPR. Two main restrictions to the predictive use of the SWATRE-CROPR combined model are that the leaf area index (LAI) and rooting depth as functions of time are a measured input. If the variation of these plant parameters should be simulated as a function of the soil water stress the combined water budget-crop growth model could be used to optimize irrigation scheduling

This paper describes a subroutine for simulating the daily extension of LAI with respect to soil water stress and a subroutine for calculating rooting depth. Both subroutines are developed for winter wheat. They are linked to SWATRE and daily values of the actual LAI and rooting depth are generated in an interactive way. The subroutines were evaluated against experimental data of winter wheat monitored in Hélécine (Belgium) and Ghazvin (Iran).

1. INTRODUCTION

Modelling of plant growth in relation to environment has increased greatly in the past few years. Particularly many attempts have been made to predict the influence of water on growth and yield under rainfed and irrigated conditions (Hanks, 1974; Rasmussen and Hanks, 1978; Feddes et al., 1978; Feyen and Van Aelst, 1983; Ritchie and Otter, 1984). These models, after having been validated can be useful in scheduling irrigation, in land evaluation studies, in assessing the benefit of water management schemes, and for other applications. However, in many cases practical application fails due to lack of some of the required input data.

The purpose of this paper is to describe a predictive model, simulating the growth and yield of winter wheat, which has the advantages of the reliable water balance approach used in SWATRE (Belmans et al., 1983), affords prediction of the leaf area index (LAI) as a function of the soil water status, calculation of rooting depth with respect to time and determination of the maximum depth of root penetration. The submodels for predicting LAI and generating the rooting depth were incorporated in SWATRE. To distinguish the extended SWATRE version from previous versions its acronym was modified to SWLEWW. The daily output of SWLEWW, the actual daily crop transpiration and the actual LAI were used as input for CROPR (Feddes et al., 1978). In this study CROPR was made specific for winter wheat by introducing appropriate values for the different model parameters. The name CROPR was therefore changed to WTGRO to indicate that the version used in this paper is specific for winter wheat.

The combined SWLEWW-WTGRO model has been verified against experimental data of growth and yield of winter wheat, cultivated under non-stressed (Hélécine, Belgium) and stressed (Ghazvin, Iran) water conditions. The model has now been developed to the point at which it can simulate the growth of a healthy winter wheat crop with adequate nutrient supply under different conditions of climate and water availability.

2. SIMULATION MODELS

The SWLEWW-WTGRO model consists of two main models (i) SWLEWW which is a field water balance model, simulating actual daily soil evaporation (E_{act}), actual daily plant transpiration (T_{act}), actual leaf area index (LAI_{act}), potential LAI (LAI_{pot}) and rooting depth and (ii) WTGRO which is

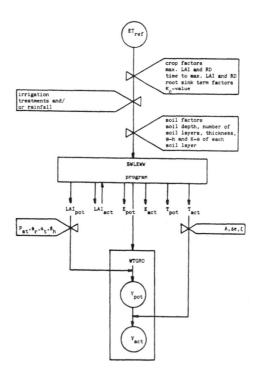

Fig. 1   Diagram representing the interactions between the two model sections, SWLEWW and WTGRO, if operating simultaneously

a crop growth model simulating the potential and actual growth, potential and actual final yield. Both models can be used independent of each other if the necessary input data are available. Linkage between the two models when used simultaneously is shown diagrammatically in Figure 1. As illustrated in this figure, daily values of rooting depth and LAI, which were hitherto inputs for SWATRE are simulated in two subroutines of SWLEWW. LAI is modelled under non-limiting water conditions (LAI_{pot}) as a function of available water (LAI_{act}). The actual LAI, generated in SWLEWW is used by this model section as input for the next day for splitting evapotranspiration into evaporation and transpiration. The potential LAI is used in WTGRO for the prediction of the potential plant growth.

Root growth and water extraction

The root water extraction rate is calculated according to Hoogland's sink term (Hoogland et al., 1981):

$$S(h,z) = \alpha(h)S_{max}(z) \tag{1}$$

and

$$S_{max}(z) = a - b(z) \tag{2}$$

in which $S(h,z)$ is the actual root water extraction rate in day^{-1} which depends on depth (z), and soil water pressure head (h), $\alpha(h)$ is a dimensionless coefficient varying from 0 to 1, depending on soil water pressure head and $S_{max}(z)$ is maximum root extraction rate decreasing with depth to account for the effect of aeration, decreasing root density, soil temperature and xylem resistance to root water uptake in day^{-1}. The values for the variables a and b in eq. (2) were assumed constant and equal to $0.01 < a < 0.03$ (day^{-1}) and b = 0 until more measured root water uptake data became available. In the present study the values of a and b have been selected in relation to type of crop. Crop type has a crucial effect on both of them and on soil water stress. This is due to the effect of crop type on rooting density and as such on the value of a. Two ranges are considered for parameter a. If water is severely limiting it is assumed that it varies between 0.005 and 0.015, whereas when soil water is amply available it varies between 0.015 and 0.030. For the crop considered a = 0.010 was taken under water stress conditions and a = 0.020 when water was not limiting. The value of b was taken constant and equal to 0.0008 whatever the state of the soil water. As a result the root water extraction rate calculated with eq. (1) is controlled by soil water through the variable α(h) which is directly dependent on soil water pressure head and through the parameter a which translates the effect of soil water stress into rooting density.

The constraint holding for eq. (1) is:

$$\int_{Z}^{0} S(h,z)dz < T_{pot} \tag{3}$$

where Z is the rooting depth in cm. Whereas Z was an input in SWATRE, here it is being simulated as a function of time and maximum root depth. According to Rasmussen and Hanks (1978) the root depth Z can be predicted as:

$$Z = DP + (DMA - DP)/\left[1 + e^{(FA - FB.DA/RTM)}\right] \tag{4}$$

in which Z is the daily value of the rooting depth in cm, DP planting depth (cm), DMA the estimated value for maximum effective rooting depth, DA the day number since planting, RTM the length of the period in days from planting until the roots reach maximum effective depth, FA and FB empirical constants. Here the value of 5 and 7 is used respectively. Figure 2 shows the simulated rooting depth for the experiments evaluated in Hélécine (Belgium) and Ghazvin (Iran).

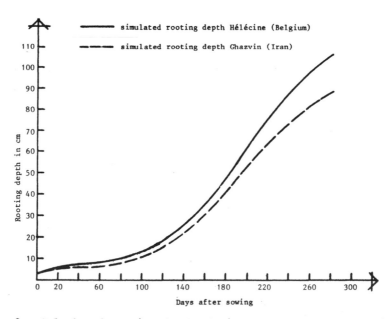

Fig. 2 Calculated rooting depth of winter wheat for the growing conditions in Hélécine (Belgium) and Ghazvin (Iran) respectively

Simulation of LAI

The LAI, expressed as the green leaf area per unit area of ground surface is simulated in the subroutine CLAI of SWLEWW. Daily values of the
actual LAI are generated as a function of soil water stress, the latter being quantified by the ratio between actual and potential transpiration rate.
The subroutine also calculates the potential increasing or decreasing rate
of LAI extension by putting the actual transpiration rate equal to the potential rate.

The equations used for generating the LAI rate of extension are different for the three growth stages into which the growing season of winter
wheat can be divided. For growth stage I, which stretches from sowing to
appreciable growth, just after the winter, it is assumed that LAI increases
linearly with time. The actual time variation of the LAI is irregular, increasing, respectively decreasing slowly with varying intensity due to winter weather conditions. The total increase of LAI in stage I is small and
of minor importance. Growth stage II starts from appreciable growth and
lasts until anthesis. According to Hanks and Puckridge (1980), eq. (5) can
be used for the simulation of the LAI extension rate during this stage of
growth:

$$LAI_{inc.rate} = \sum_{D=DB}^{DLM} \left[Acos(A.D - B)LM.RU/2 \right] T_{act}/T_{pot} \tag{5}$$

in which $LAI_{inc.rate}$ is the increasing rate of LAI, D is time increment in
days, DB and DLM are the day number at the beginning and end of stage II
respectively, reckoned from the sowing date, LM is the maximum obtainable
LAI, T_{act} and T_{pot} are the actual and potential transpiration in $mm \cdot d^{-1}$
respectively, A and B are constants which can be calculated from:

$$A = \pi/(DLM - DB) \tag{6}$$

and

$$B = A.DLM - \pi/2 \tag{7}$$

In eq. (5) soil water stress is accounted for by the ratio T_{act}/T_{pot}.
When water is not limiting growth the subroutine by-passes the ratio
T_{act}/T_{pot} and simulates potential daily rates of LAI extension. The RU coef-
ficient in eq. (5) takes into consideration the effect of plant density on
LAI. Its value is equal to 1 for a planting density of 90 kg seed per ha.
RU is taken greater or smaller for seed rates above or below 90 kg per ha.

117

The effect of planting density on RU is considered when LAI is less than
1.5. Thereafter, due to the compensatory effect of tillering RU is kept constant and equal to 1.

In stage III, which stretches from anthesis to maturity, LAI decreases
and the decreasing rate can be calculated according to eq. (8) (Hanks and
Puckridge, 1980):

$$LAI_{dec.rate} = \sum_{SET_{ref}(DLM)}^{SET_{ref}(DMA)} C\cos(C.SET_{ref} - D)LH.ET_{ref}/2 \tag{8}$$

in which $SET_{ref}(DLM)$ and $SET_{ref}(DMA)$ are the cumulative potential evapotranspiration of the reference crop grass (Burman et al., 1981) in mm
at the day of LAI anthesis and of maturity respectively, both reckoned
from DB, the 1st day of stage II. ET_{ref} in $mm \cdot d^{-1}$ is the daily value of the
potential evapotranspiration rate of a full grown grass, LH is the maximum
LAI calculated by eq. (5), C and D are constants equal to:

$$C = \pi / \left[SET_{ref}(DMA) - SET_{ref}(DLM) \right] \tag{9}$$

and

$$D = C.SET_{ref}(DLM) - \pi/2 \tag{10}$$

If for evaluation purposes no measured data of the LAI are available,
predicted values of LAI can be converted into ground cover fraction data
using the approach suggested by Adams and Askin (1977) and use the observed
GC fraction data for validation. For LAI values smaller than 3.5 the following equation can be used:

$$GC = 1 - e^{-K.LAI} \tag{11}$$

in which GC is the daily fraction of the ground cover and K is a dimensionless light-extinction coefficient. In this study a value of 0.6 was used
for K. For LAI values equal or greater than 3.5 it is assumed that the crop
covers the soil surface completely. Under this condition the GC fraction is
taken equal to 1 for the rest of the growing season.

The predicted value of the GC fraction is used in SWLEWW the next day
for separating potential crop evapotranspiration ($= K_c.ET_{ref}$) into soil
evaporation and plant transpiration. For GC < 0.45 K_c is taken equal to 1,
for GC $\geqslant$ 0.45 the value K_c is selected according to Doorenbos and Pruitt
(1977).

WTGRO model

WTGRO differs from CROPR in the way that the temperature effect on growth is accounted and the replacement of GC fraction by the fraction of incoming photosynthetically active radiation (IPAR) in the equation for calculating the daily potential growth (q_{pot} in kg DLM·ha^{-1}·d^{-1}).

According to Hodges and Kanemasu (1977) IPAR can be estimated from the LAI which results in a more appropriate input for the estimation of q_{pot}:

$$IPAR = 0.5739 \ LAI^{0.3296} \tag{12}$$

To correct the rate of photosynthesis of winter wheat for temperature effects the following relationship was used (Rickman et al., 1975):

$$\alpha_t = e^{-\beta \left[(TEMP_{opt} - TEMP)/TEMP \right]^2} \tag{13}$$

where α_t is a temperature correction factor varying between 0 and 1, $TEMP_{opt}$ is the optimum air temperature for growth of winter wheat reported to vary from 15 to 25°C (Friend, 1966; Macdowall, 1973) and β is the low temperature cut-off point for growth to be derived from field observations. Even at temperatures of 0°C small growth rates have been reported for wheat plants (Van Dobben, 1962). TEMP is the mean daily value of the air temperature. All temperatures for the calculation of the temperature correction factor α_t are in °K. The value of β and $TEMP_{opt}$ are variety dependent. In this study $TEMP_{opt}$ was taken 293.15°K for the Ghazvin experiment. The value of the parameter β was derived from the assumption that $\alpha_t = 0.0005$ for TEMP = 273.15°K. Inserting these values in eq. (13) yields a value $\beta = 1418$. The value of $TEMP_{opt} = 293.15$°K and the calculated value for β of 1418 were used to calculate the value of α_t as a function of the daily average air temperature. In the case of Hélécine $TEMP_{opt}$, and α_t were taken 288.15°K and 0.02 respectively and a value of 1297 was derived for β.

The potential growth rate (q_{pot} in kg DLM·ha^{-1}·d^{-1}) is calculated as follows:

$$q_{pot} = P_{st} \times IPAR \times \phi_r \times \alpha_t \times \beta_h \tag{14}$$

in which P_{st} is the gross potential growth rate of a standard canopy expressed in kg carbyhydrate per ha and per day and calculated according to the photosynthesis submodel of De Wit (1965). Coefficient ϕ_r is the respiration and maintenance factor which reduces the gross potential rate to the net potential growth rate. For the moderate climate of Hélécine a value of

0.75 (Sibma, 1977) was taken, whereas for the semi-arid climate of Ghazvin it was taken equal to 0.53 (Hodges and Kanemasu, 1977). The harvest index β_h reduces the net potential growth rate of the total dry matter to the fraction of the grain yield. Throughout the growing season β_h was taken equal to 1. To calculate grain yield from total dry matter a value of 0.40 (Sibma, 1977; Slabbers and Dunin, 1981) was assigned to β_h for both regions.

The actual daily growth rate of winter wheat as a function of the water availability is calculated according to Feddes et al. (1978) using daily values of q_{pot}, T_{act} and Δe (= water vapour pressure deficit in mbar) as input. In addition, to use this approach the following parameters have to be defined: A, the water use efficiency factor (kg $DLM \cdot ha^{-1} \cdot mm^{-1} \cdot mbar$) and the mathematical flexibility factor ξ, which is dimensionless and stands for the effect of non-controllable influences on yield. In fact ξ can be considered as a matching factor to tune the simulated results to field measurements. A value of 490 kg $DLM \cdot ha^{-1} \cdot mm^{-1} \cdot mbar$ was given to A and values of 0.1 and 0.001 were assigned to ξ for Ghazvin and Hélécine respectively.

3. EXPERIMENTAL DATA

The data used to evaluate the performance of SWLEWW-WTGRO came from two wheat experiments run under completely different climatic conditions. One of the experiments was situated in Hélécine (Belgium) characterized by a moderate humid climate and the other one was located in Ghazvin (Iran), having a semi-arid climate. Under normal climatic conditions water stress is rather exceptional for winter wheat in Belgium grown on a deep loam, whereas in the Ghazvin area irrigation of wheat is a common agricultural practice.

In Ghazvin the experiment was run during the 1975-76 growing season, the main objective being to evaluate the growth response of the winter wheat variety OMID to irrigation and fertilization. The experimental site was situated 150 km west of Tehran (36°15' N, 50° E and located 1300 m above mean sea level). The experiment was arranged in a factorial design with three irrigation and three fertilizer levels. The irrigation treatments were based on class A pan evaporation measurements. Details of the irrigation treatments applied are given in Tables 1 and 2.

The fertilizer treatments applied were respectively F1 = 120 N, 90 P, 40 K in $kg \cdot ha^{-1}$, F2 = 90 N, 60 P, 30 K and F3 = 60 N, 40 P, 20 K. Each of the irrigation-fertilizer treatments were replicated three times. The soil

TABLE 1   Time schedule and irrigation doses for the irrigation treatments applied to the winter wheat experiment in the Ghazvin area (Iran), during the 1975-76 growing season

Number of irrigation application	1	2	3	4	5
Irrigation date	13.10.75	7.11.75	21.04.76	25.05.76	15.06.76

Irrigation treatment	Irrigation dose in mm				
I1	118	100	98.0	90.0	82.0
I2	118	100	73.5	67.5	61.5
I3	118	100	49.5	45.0	41.0

TABLE 2   Totals of irrigation and rainfall per irrigation treatment for the winter wheat experiment in the Ghazvin area (Iran) during the 1975-76 growing season

Irrigation treatment	Total irrigation water applied in mm	Total rainfall in mm	Total water in mm
I1	488.0	223	711.0
I2	420.5	223	643.0
I3	353.5	223	576.5

of the experimental site was loamy sand containing 130 mm of water per meter of soil depth. The soil profile was characterized physically and the θ-h and K-θ relationships per layer were used as input for SWLEWW. The climatological data needed to run both models were obtained from a nearby weather station 10 km away. Daily values of the necessary climatic variables were available. The monthly average values of these variables and the cumulative monthly rainfall for the growing season considered are summarized in Table 3.

The crop was sown on October 10, 1975 and harvested on July 4, 1976. Only the crop data monitored on the highest fertilizer treatments were used for validation purposes to meet the condition of adequate nutrient supply.

The Hélécine experimental site is situated in the loamy region of Belgium (51°50' N, 5° E and 53 m above mean sea level). The soil has a total available water depth of 230 mm in 1 m soil depth. On November 2, 1982 the winter wheat variety CORIN was sown and harvested on August 8. 1983. Six nitrogen fertilizer treatments differing in total amount or dosage were ap-

TABLE 3   Monthly average values of the daily mean air temperature (°C), the daily mean relative humidity (%), the daily mean wind speed (m·s^{-1}) at 2 m height, the cloudiness (oktas), the calculated daily reference evapotranspiration (mm·d^{-1}) and the cumulative monthly rainfall (mm) at Ghazvin (Iran) for 1975-76

Month	Mean TEMP in °C	Mean RH in %	Mean WS in m·s	Cloudiness in oktas	ET$_{ref}$ in mm·d^{-1}	Cum. rainfall in mm
October	12.3	50.1	0.8	1.8	2.8	0.0
November	6.2	58.5	1.3	4.0	1.5	15.1
December	0.0	72.3	1.0	5.2	0.8	37.6
January	1.3	67.8	1.0	3.7	1.1	11.4
February	-1.3	70.7	1.2	4.5	1.3	21.5
March	1.6	69.2	1.2	4.5	2.1	64.3
April	10.9	62.8	1.6	5.4	3.2	26.8
May	15.5	62.4	1.3	4.6	4.5	40.3
June	23.1	48.3	1.5	1.7	7.2	6.0
July	28.3	49.3	1.4	1.5	7.2	0.0

plied on the experimental field with three replications. The crop was completely dependent on rainfall. Again the crop data monitored on the optimal nitrogen fertilizer treatment were chosen for validation purposes. It was assumed that on this treatment crop growth was not affected by soil fertility. All the necessary climatological variables needed to run SWLEWW-WTGRO were measured on the experimental site. The soil profile has been characterized chemically and physically. The LAI was measured 5 times throughout the growing season. On each measuring date the LAI of 20 complete wheat plants, collected at random within each observation plot, were measured with a LI-30501 area meter. In Ghazvin only the ground cover fraction was monitored 4 times throughout the growing season.

4. RESULTS AND DISCUSSION

In Figure 2 the extension of root depth versus time is given for the winter wheat experiments in Ghazvin and Hélécine. The maximum calculated effective root depths were 88 cm and 106 cm respectively. For Ghazvin no field data were available to validate this figure. For the area it is however generally accepted that the roots of winter wheat penetrate no deeper than 80 to 90 cm. The maximum root depth of 106 cm as derived for Hélécine corresponds well with profile pit observations. Not enough field data on both experimental sites were available to validate the generated time variation of root depth. According to Belmans et al. (1979) the pattern of root growth and maximum rooting depth is controlled by the soil water status, soil aera-

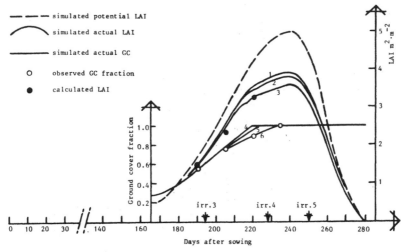

Fig. 3   Simulated and observed actual ground cover fraction values for
winter wheat grown in 1975-76 in Ghazvin, Iran. The dashed line gives
on a day-to-day basis the simulated potential LAI. Lines 1, 2 and 3
represent the actual simulated LAI for irrigation treatment I1, I2 and
I3 respectively. Lines 4, 5 and 6 give the corresponding derived ground
cover fraction

tion, soil mechanical resistance, soil temperature, distribution of solutes
(both nutritive and toxic), etc. Here maximum rooting depth is being given
as input, which weakens the predictive ability of this section. Therefore
some more research and field observations should be carried out e.g. to extend eq. (4) in such a way that root penetration could also be simulated
as a function of the water status in the root zone. For the time being
maximum rooting depth has to be derived from local field experiments.

The dashed line in Figure 3 shows the time course of the potential LAI
for winter wheat, as simulated for Ghazvin farming practices. Lines 1, 2
and 3 show the variation of the actual LAI, calculated for irrigation treatments 1, 2 and 3 respectively (see Tables 1 and 2). The points around curve
3 represent LAI values that have been calculated from ground cover fraction
data using eq. (11). Lines 4, 5 and 6 in Figure 3 show the corresponding
ground cover fraction. As can be seen line 6, showing the variation of the
calculated ground cover fraction with time for irrigation treatment 3,
matches fairly well the observed ground cover fraction data.

Figure 4 shows the measured and simulated values of the LAI of the crop
grown in Hélécine. The open circles represent the observed values of the
leaf area index. The full line represents the actual simulated LAI on a
day-to-day basis, while the dashed line shows the time variation of the cal-

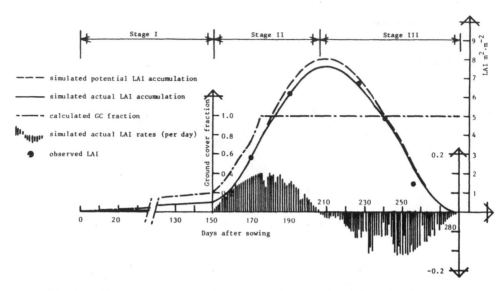

Fig. 4   Simulated and measured actual LAI for winter wheat grown in a
moderate climate (Hélécine, Belgium). The figure shows also on a day-to-
day basis the calculated potential LAI, the actual simulated LAI-rate
and the estimated ground cover fraction

culated, potential LAI. The vertical line segments from day 150 after sowing until harvesting date represent the daily simulated values of the LAI
extension rate. The maximum LAI-value of 8, reached by the crop at the end
of growth stage II is an input to SWLEWW. In addition to the maximum obtainable potential value of LAI the dates at the beginning and end of each
growth stage have to be given as input. The increase in the actual and potential LAI during stage II is entirely time dependent. The reduction from
potential to actual level is controlled solely by the ratio between actual
and potential transpiration. As can be observed from Figure 4 the transpiration was slightly depressed by water stress. The agreement between measured
and simulated LAI under the non-stressed conditions in Hélécine is satisfactory. For completeness the variation of the ground cover fraction with time
(the dot-dashed line) has been given in Figure 4 as well. When LAI reaches
the value of 3.5 the GC fraction is put equal to 1 and kept unchanged for
the rest of the growing season.

The introduction of a crop phenology model will have the advantage that
the dates at the beginning and end of the different growth stages would be
predicted instead of being supplied as an a priori input. According to Weir
et al. (1984) it is feasible to simulate the phenological development of
winter wheat as a function of temperature, day length and vernalization. By

124

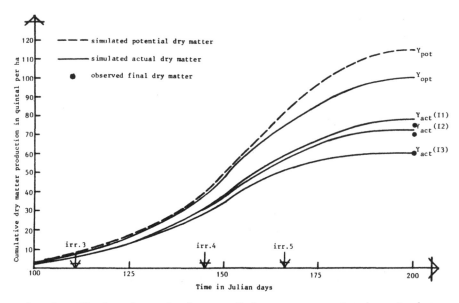

Fig. 5   Simulated cumulative total dry matter production of winter
wheat in quintal per ha at Ghazvin (Iran) under potential, optimal and
actual growing conditions. The actual cumulative total dry matter production is simulated for each of the irrigation treatments applied

doing so the model approach will become more predictive, but on the other
hand the number of parameters will increase as well as the complexity of
the procedure.

Calculated and measured dry matter accumulation of winter wheat for the
growing seasons analysed are shown in Figures 5 and 6. Figure 5 summarizes
the simulation runs for the Ghazvin area for the 1975-76 growing season,
while Figure 6 gives the same information for the 1982-83 growing season in
Hélécine. For the Ghazvin experiment only the final actual yield of total
dry matter for the three observed irrigation levels were available for validation. In Hélécine, however, the data on the total dry matter of 9 interim harvests, distributed almost uniformly over the growing season, were
available for testing the performance of SWLEWW-WTGRO. The measured dry
matter data in both locations were monitored on plots with the maximum fertilization level, so as to meet the model assumption of no nutrient limitation. The model predicts the accumulation of dry matter of winter wheat as
a function of water availability. In Table 4 the calculated potential and
actual evaporation, transpiration and evapotranspiration for the treatments
analyzed are given. The relative transpiration can be regarded as a cumulative measure of the average water status throughout the growing season.

125

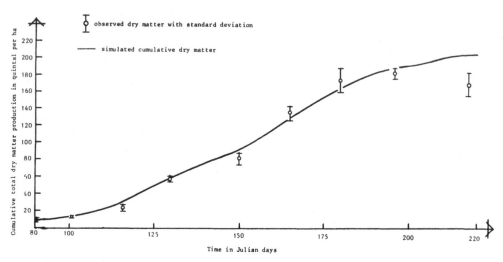

Fig. 6   Simulated and observed cumulative total dry matter production of winter wheat in quintal per ha at Hélécine (Belgium)

TABLE 4   Calculated cumulative actual and potential evaporation, transpiration, evapotranspiration and relative transpiration of winter wheat for the growing season 1982-83 in Ghazvin and the growing season 1975-76 in Hélécine

Location and treatment	Cumulative evaporation in mm		Cumulative transpiration in mm		Cumulative evaporation in mm		Relative transpiration
	E_{pot}	E_{act}	T_{pot}	T_{act}	ET_{pot}	ET_{act}	T_{act}/T_{pot}
Ghazvin, I1	259.6	193.7	482.4	360.1	742	553.8	0.75
Ghazvin, I2	260.3	194.4	481.7	327.5	742	521.9	0.68
Ghazvin, I3	268.8	202.9	473.2	257.9	742	460.8	0.55
Hélécine	120.5	115.5	367.2	329.0	487.7	444.5	0.90

In Figure 5 the actually observed final dry matter for irrigation treatments I1, I2 and I3 were 7392.5, 6987.5 and 6092.5 kg·ha^{-1} respectively. The simulated final total dry matter in kg per ha were 7750, 7260 and 5990 for the wettest to the driest irrigation treatment. If in the simulation approach it is assumed that without water limitations actual crop transpiration equals potential transpiration for Ghazvin growing conditions an optimal total dry matter yield (Y_{opt}) of 10 000 kg DM·ha^{-1} is obtained. Comparing the ratio T_{act}/T_{pot} with Y_{act}/Y_{pot} for the three irrigation levels examined a more or less linear relation is obtained. In the simulation of Y_{act} and Y_{opt} it is assumed that the flexibility factor has a value of 0.1, indicating that apart from water other growth factors e.g. nutrient deficiency, diseases and damage due to hot winds, hamper dry matter accumulation. By putting the flexibility factor equal to 0 and assuming water is not limiting an estimate of the potential dry matter accumulation of winter wheat for the Ghazvin area is obtained. In this case a value of 11 509 kg DM·ha^{-1} was found for Y_{pot}. The discrepancy between total simulated and measured dry matter yield ranges from 1.7 to 4.8%.

The data derived for Hélécine and presented in Figure 6, show excellent agreement between predicted and measured dry matter accumulation. The discrepancy between predicted and final measured dry matter can be attributed to sampling losses. To get this agreement ξ was given a value of 0.001 which implies that in the Hélécine experiment the effect of growth limiting factors other than soil water stress is less than for the Ghazvin experiment where ξ was 0.1.

5. CONCLUSIONS

In general the SWLEWW-WTGRO model appears to give accurate prediction of dry matter yield as influenced by water stress. The advantage of the given approach in comparison to the SWATRE-CROPR procedure is that the extension of the green leaf area index is generated internally as a function of water status. If measured values of LAI are not available for validation or evaluation purposes the model converts simulated LAI values into ground cover fraction. It is suggested that in the present state of the model the harvest index should be kept constant and equal to 1. Further improvement of the described model should be oriented towards the incorporation of a phenological submodel and a subroutine that allows the prediction of rooting depth as a function of soil water and other related factors. At present SWLEWW-WTGRO can be used under quite differing climatic conditions for predicting the effect of irrigation on dry matter yield when nutrients are not limiting.

REFERENCES

Adams, J.E. and Askin, G.F. 1977. A light interception method for measuring row crop ground cover. Soil Sci. Soc. Amer. J. 41: 780-792.
Belmans, C., Feyen, J. and Hillel, D. 1979. An attempt at experimental validation of macroscopic-scale models of soil moisture extraction by roots. Soil Sci. 127: 174-187.
Belmans, C., Wesseling, J.G. and Feddes, R.A. 1983. Simulation model of a water balance of a cropped soil: SWATRE. J. Hydrol. 63: 271-286.
Burman, R.D., Nixon, P.R., Wright, J.L. and Pruitt, W.O. 1981. Water requirements. In 'Design and operation of farm irrigation systems' (ed. M.E. Jensen). ASAE Monograph 3: 189-232.

Doorenbos, J. and Pruitt, W.O. 1977. Crop water requirements. FAO Irrig.
and Drain. Paper 24. 156 pp.

Feddes, R.A., Kowalik, P.J. and Zaradny, H. 1978. Simulation of field water
use and crop yield. Simulation Monograph, PUDOC, Wageningen. 189 pp.

Feyen, J. and Van Aelst, P. 1983. Test of a simple growth model simulating
sugar beet yields. Pedologie 33(3): 281.293.

Friend, D.J.C. 1966. The effects of light and temperature on the growth of
cereals. In 'The growth of cereals and grasses' (eds. F.L. Milthorpe
and J.D. Ivins). Butterworth Inc., Washington: 181-199.

Hanks, R.J. 1974. Model for predicting plant yield as influenced by water
use. Agron. J. 66: 660-665.

Hanks, R.J. and Puckridge, D.W. 1980. Prediction of the influence of water,
sowing date and planting density on dry matter production of wheat.
Aust. J. Agric. Res. 31: 1-11.

Hodges, T. and Kanemasu, E.T. 1977. Modeling daily dry matter production of
winter wheat. Agron. J. 69: 974-979.

Hoogland, J.C., Belmans, C. and Feddes, R.A. 1981. Root water uptake model
depending on soil water pressure head and maximum extraction rate. Acta
Hort. 119: 123-136.

Macdowall, F.D.H. 1973. Growth kinetics of marquis wheat. IV. Temperature
dependance. Can. J. Bot. 51: 729-736,

Rasmussen, V.P. and Hanks, R.J. 1978. Spring wheat yield model for limited
moisture conditions. Agron. J. 70: 940-944.

Rickman, R.W., Ramig, R.E. and Allmaras, R.R. 1975. Modeling dry matter accumulation in dryland winter wheat. Agron. J. 67: 283-290.

Ritchie, J.T. and Otter, S. 1984. Description and performance of CERES-
wheat: a user oriented wheat yield model. Report of the USDA, Agricultural Research Service. 27 pp.

Sibma, L. 1977. Maximization of arable crop yields in the Netherlands. Neth.
J. Agric. Sci. 25: 278-287.

Slabbers, P.J. and Dunin, F.X. 1981. Wheat yield estimation in Northwest
Iran. Agric. Water Manag. 3: 291-304.

Van Dobben, W.H. 1962. Influence of temperature and light condition on dry
matter distribution, development rates and yield in arable crops. Neth.
J. Agric. Sci. 10: 377-389.

Weir, A.H., Bragg, P.L., Porter, J.R. and Rayner, J.H. 1984. A winter wheat
crop simulation model without water or nutrient limitations. J. Agric.
Sci., Camb. 102: 371-382.

Wit, C.T. de. 1965. Photosynthesis of leaf canopies. Agric. Res. Rep. 663.
PUDOC, Wageningen. 57 pp.

Water balance and crop production

H.C.ASLYNG
Hydrotechnical Laboratory, RVAU, Copenhagen, Denmark

ABSTRACT

The interrelationships between climate, soil and plant have in simplified form been combined in the model WATCROS (WATer balance and CROp production Simulation). The evapotranspiration is based on the Penman or the
Makkink equation. Plant development is based on temperature sum. Gross and
net plant production are predicted from efficiency of absorbed photosynthetic active light and plant respiration, based on crop species, crop mass
and temperature.

Of the climatic factors global radiation, air temperature and precipitation, the latter is the variable which changes most with time and space.
Therefore it is necessary to record precipitation locally and at ground
surface. Critical parameters in the water balance are root growth, effective root depth and soil available water. In crop production the respiration is the most critical parameter.

In Denmark half of the agricultural land is sandy and need irrigation
as the root zone capacity for available water is low. About 25% of the sandy soil is under sprinkler irrigation. On an average the actual crop yield
is, depending on soil and region, 40 to 55% of the potential yield. Of winter wheat e.g. the potential yield is 10-13 t grain/ha (15% water).

1. INTRODUCTION

The WATer balance and CROp production Simulation model WATCROS is described by Aslyng and Hansen (1982, 1985). It is an EDP implemented mathematical model whose main aim is to simulate water balance (evapotranspiration, deep percolation and change in soil water storage) and potential and
water limited production of agricultural crops. As a basic assumption soil
pH, drainage, tillage, structure, nutrition, crop stand, diseases and pests
are considered non-constraining.

The crop and the soil are characterized by a number of model parameters.
The model includes standard values of these parameters but they may be
changed by option. The climate is characterized by daily values of global
radiation, air temperature, precipitation and potential evapotranspiration.
The potential evapotranspiration can be estimated by the Penman (1956)
equation but may also be estimated by the Makkink (1967) equation (Aslyng
and Hansen, 1982 and 1985) and then only the variables global radiation,
air temperature and precipitation are required.

2. ROOT ZONE CAPACITY

The root zone capacity depends on the soil capacity for plant available water, which may be known, and the effective root depth, the latter being defined as the soil depth to which the root density is at least 0.1 cm root per cm^3 soil. In the beginning of the growth period the root zone capacity increases due to root growth. The effective normal root growth per day is assumed to be 1.2 cm for grass and 1.5 cm for other crops. Root growth is assumed to begin at the onset of plant growth and to stop when maximum effective rooting depth has been reached. The maximum depth depends on plant genetic factors and soil properties.

3. GROWTH PERIOD AND CROP SURFACE DEVELOPMENT

The crop development is based on the sum of the daily (24 h) temperature from 1 March (Table 1). For grass and fodder beets the end of the growth pe-

TABLE 1   Temperature sum and growth period for different crops

Crop	Temperature sum from 1 March				Average growth period		
	Onset growth	Crop cover*	Ripening		Onset	End	Days
			Onset	End			
W. wheat	125	425	1425	1750	15.4	15.8	125
W. barley	100	350	1000	1325	10.4	20.7	100
S. barley	225	625	1300	1625	1.5	10.8	100
S. rape	300	700	–	1900	10.5	25.8	110
F. beets	550	1450	–	–	1.6	30.10	155
Grass	125	425	–	–	15.4	30.10	200

*Crop Area Index (CAI) >5

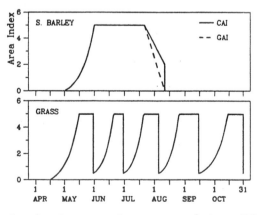

Fig. 1   Model for development of crop area index, CAI, and green area index, GAI

130

riod is assumed to be 31 October. The development of the crop surface is
characterized by model estimated crop area index, CAI, and is illustrated
in Figure 1, where summer barley and grass are shown as examples. The
criteria for cutting grass are:

Cut number	1	2	3	4	5
DM, $t \cdot ha^{-1}$	5	6	6	5	5
Days since previous cut/onset	45	35	35	40	75

4. EVAPOTRANSPIRATION

Input to the model is precipitation P, irrigation I and potential evapotranspiration E_p. Vegetation is characterized by CAI and effective root
depth. The soil is characterized by the amount of water which readily can
be evaporated from the soil surface, and by the amount of water available
to the plants in the root zone, i.e. the top soil capacity and the root
zone capacity.

An a priori assumption is that the actual evapotranspiration, E_a, can
reach but not exceed the potential evapotranspiration. The E_a/E_p ratio
varies during the day and season depending on the evaporative demand and
the availability of soil water. The fraction X of the available water capacity which can be utilized at potential level is a function of the season.
The value of X is assumed to be 0.6 in April, 0.5 in May-August, 0.6 in
September and 0.7 in October. Winter wheat is known for its capacity for
osmotic adjustment and efficiency in extracting moisture from the soil. The
X-value therefore is assumed to be 0.1 larger in the case of winter wheat.

The potential evapotranspiration is divided between the soil and the
crop based on crop development, net radiation and Beer's law. The extinc-
tion coefficient K is chosen equal to 0.6. If neither the crop nor the soil
can fulfill the evaporative demand, the energy is not used for evaporation.

The interception storage is 0.5 CAI but max. 2.5 mm. Not intercepted,
P or I is supplied to the top soil and the root zone reservoir. The capacity of the top soil reservoir is assumed independent of the soil type and
crop species and is stated to be 10 mm. Water in surplus of the interception capacity, the top soil storage capacity and the root zone storage
capacity is allocated to a through-flow reservoir where the surplus remains
for 3 days. If not evapotranspirated during that period the surplus is
drained out of the root zone as deep percolation.

A special case occurs when the root zone water storage has been ex-

hausted to more than the fraction X and the top soil is rewetted to field
capacity. The vegetation will then primarily use this water as source for
transpiration until the soil is exhausted to the level of X before the whole
whole root zone again is considered.

5. GROSS AND NET PRODUCTION

The potential gross production is predicted from crop development, ab-
sorption of net photosynthetically active radiation (PAR) and photosynthetic
efficiency. The efficiency is lower at high than at low PAR and is on an
average assumed to be 8% for April to August, 9% for September and 10.5%
for October. Potential net production is predicted by subtracting the estimated respiration from the predicted gross production. The daily maintenance respiration, M, is equivalent to a fraction of the total crop dry
matter and the growth respiration, G, is a 'constant'. On basis of literature studies the values given in Table 2 have been selected. The change for
winter wheat takes place in the model at a temperature sum of 1000. For M
a Q_{10} = 2 is adopted.

TABLE 2   Maintenance, M, and growth, G, respiration, M/day

Respiration	Winter wheat*	Winter barley	Spring barley	Spring rape	Fodder beet	Rye grass
M (r_m), %	1.5/1.0	1.5	1.5	2.0	0.8	4.0
G (r_g), %	30/20	30	30	30	30	30

*the first figure for the pre-anthesis and the second figure for the
post-anthesis phase (Van Keulen and De Milliano, 1984)

Water limited gross production is estimated as potential gross production reduced by use of the factor E_a/E_p. The respiration and net production
are calculated in the same way as for the potential case.

It is assumed that all harvested material is removed from the field.
Only roots, stubble and unavoidable losses are left as residue. The quantity of residue is assumed to be 3 t DM/ha with normal farm harvesting but
with experimental harvesting the quantity depends on crop species (Table 3).
In the case of grass the residue is assumed to amount to 3 t DM/ha when the
crop has reached its final root depth. If a cut is performed before the
final root depth has been reached the residue is calculated as a fraction
of the 3 t DM/ha and this fraction is determined as the ratio between the
actual root depth and the final root depth.

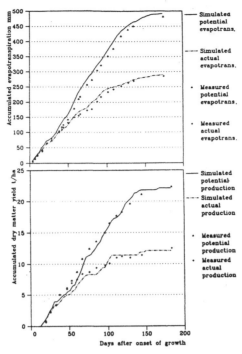

Fig. 2   Accumulated evapotranspiration and yield of grass at Tåstrup
1976

TABLE 3   Residue in the field after different crops, nitrogen fertil-
izations and harvest methods, t DM and kg N/ha

| | Farm harvest | | | | Experimental harvest | | | |
| | -N | | +N | | -N | | +N | |
	DM	N	DM	N	DM	N	DM	N
Cereal	3	35	3	50	2	30	2	35
Rape	3	35	3	50	2	30	2	35
Beet	3	35	3	50	1	20	1	20
Grass*	3	50	3	50	3	50	3	50

6. EXPERIMENTAL AND PREDICTED DATA

As examples some of the obtained data are presented in the following.

1) Tåstrup (DK) 1976: grass (Lolium perenne) yield in 5 cuttings with and
without irrigation on the experimental farm Højbakkegård at the climate station at Tåstrup 1976. Soil moisture was recorded with the neutron method. The
meteorological data are from the climate station. The soil is type 6, the

133

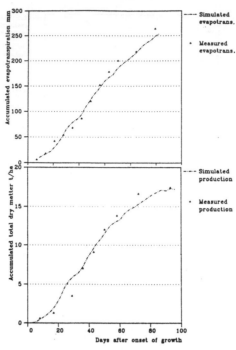

Fig. 3   Accumulated evapotranspiration and dry matter (incl. roots) of spring barley at Jersie 1979

TABLE 4   Water consumption and dry matter yield simulated and obtained in a field experiment with grass, Tåstrup 1976

Level	Evapotranspiration, mm		Dry matter yield, t/ha	
	Model	Exp.	Model	Exp.
Potential	492	481	21.7	22.3
Water limited	291	286	12.1	12.5

root growth 1.2 cm per day, the effective root depth 100 cm and the root zone capacity 170 mm (Jensen, 1980) (Figure 2).

2) Jersie (DK) 1979: barley on clay soil (7) on a farm at Jersie. The root growth was around 2 cm per day, the effective root depth 125 cm and the root zone capacity 225 mm. Precipitation and soil moisture (neutron method) were recorded. Global radiation and air temperature are from the climate station at Tåstrup (19 km northeast of Jersie). The transpiration and production were almost at the potential level without irrigation. The barley variety was Welam (Nielsen, 1980) (Figure 3).

Final evapotranspiration and yield are given in Tables 4 and 5.

134

TABLE 5   Water consumption and grain yield simulated and obtained in a field experiment with spring barley, Jersie 1979

Level	Evapotranspiration, mm		Grain (15% water), t/ha	
	Model	Exp.	Model	Exp.
Potential	312	–	8.0	–
Water limited	253	264	7.4	7.7

The agreement between simulated and experimental data presented in Figures 2 and 3 and Tables 4 and 5 is satisfactory. In the water limited case (Table 5) the experimental data are 5% larger than the simulated data.

3) Rothamsted (UK) 1979: field experiment with barley at Rothamsted Experimental Station 1979 on a clay soil. The root growth was around 2 cm per day, the effective root depth 125 cm and the root zone capacity 225 mm. Global radiation, air temperature and precipitation have been recorded at the same location where the field experiment was carried out. There was one treatment with irrigation (ample water supply) and one without irrigation and also protected against precipitation. Figure 4 shows satisfactorily good agreement between simulated and experimental yield (Personal communication with R.P. Scammell, Rothamsted).

In Figure 5 the total harvested and the predicted yields of cereals and rape with and without nitrogen fertilizer at Tåstrup 1984 are presented. The average grain yield of fertilized winter wheat was 10.5 and of unfertilized 4.5 t/ha (15% water). The yield of summer barley was 7.3 and 3.1 (15% water) and of summer rape 4.3 and 1.9 t/ha (9% water) respectively.

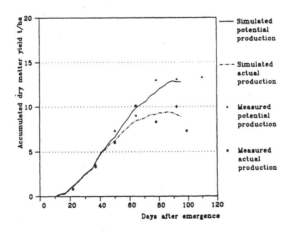

Fig. 4   Accumulated dry matter yield of spring barley at Rothamsted 1979

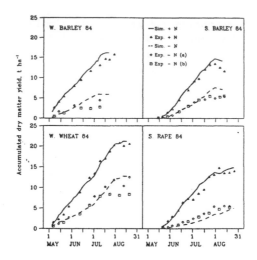

Fig. 5 Dry matter yield. In case (a) the actual and (b) also the previous crop was unfertilized. (a) and (b) yields differ for winter but not for spring crops. The simulated yield of rape comprises also DM in the shedded leaves

7. CLIMATE, SOIL AND WATER BALANCE

Climatic data from the Danish Karup, Stevns and Tåstrup regions collected over a number of years have been applied to the model WATCROS and model NITCROS. The soils, periods and geographical positions are: coarse sand (soil 1) at Karup, and clay with fine sand (soil 6) at Stevns for the 9 years 1968-1977, and clay with fine sand (soil 6) at Tåstrup for the 18 years 1966-1984. The Karup (region) is situated at $56^{\circ}16'$ N, $9^{\circ}9'$ E, the Stevns (region) at $55^{\circ}19'$ N, $12^{\circ}40'$ E, and the Tåstrup climate and experimental station at $55^{\circ}40'$ N, $12^{\circ}18'$ E.

For the stated periods and regions we have the required daily values for global radiation, air temperature, precipitation at soil surface and potential evapotranspiration. For the agricultural regions the annual averages including also actual evapotranspiration are as follows:

	Karup 1968-1977	Stevns 1968-1977	Tåstrup 1968-1982
Global radiation, MJ m^{-2}	3545	3800	3720
Temperature, $^{\circ}$C	7.4	7.5	7.5
Precipitation, mm	775	605	590
Potential evapotranspiration, mm	540	580	575
Actual evapotranspiration, mm	365	430	430
Available soil water capacity, mm	60	170	170

TABLE 6   Water balance as annual average by growing different crops at Tåstrup and Karup, mm

| | Tåstrup 1966-1984 | | | Karup 1968-1977 | | |
	Precipitation	E_a	$P-E_a$	Precipitation	E_a	$P-E_a$
Winter wheat	586	458	128	775	370	405
Summer barley		422	164		357	418
Summer rape		435	151		368	407
Fodder beets		418	168		358	417
Grass		456	130		381	394

For Stevns and Tåstrup the climatic and soil conditions do not differ much from each other. The Karup region has a larger precipitation but is nevertheless the one that suffers most from drought because of the sandy soil having a small root zone capacity.

The precipitation is in all cases very variable as to time and space. Even if the total precipitation in the growth period may be sufficient for covering the water demand for potential production, crops may suffer from water stress from time to time because of irregular distribution of rainfall.

The predicted evapotranspiration and the estimated average annual water balances for different crops without irrigation are for Tåstrup and Karup presented in Table 6. The data show that the water balances are not greatly dependent on crop species. The E_a-values are much larger and the $P-E_a$ values (net precipitation) are much smaller at Tåstrup than at Karup. The E_a-values do to some extent reflect the water supply and the difference between the regions.

In Figure 6 the predicted potential and the water limited yields of some of the crops for each of the years 1966-1984 at Tåstrup are illustrated. From total dry matter yield the estimated grain yield with 15% water is obtained by assuming a harvest index of 45 for winter wheat and of 50 for summer barley. These could be the actual indices with normal farm harvesting and spraying of winter wheat for shortening of the straw. The harvest indices have very likely increased during the considered periods. From comparison the actual yield recorded as a 'normal' yield by the nearby experimental station for the same soil type has been given too. The experimental station is Roskilde situated close to Tåstrup.

The actual yield may, in general, be lower than the water limited yield as maximum utilization of available water may not always be possible. For

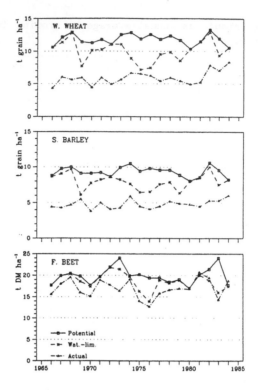

Fig. 6 Predicted potential and water limited yield at Tåstrup and recorded actual (normal) yield at Roskilde Experimental Station

some of the years water limited and actual yields are almost equal. The actual yields may have been influenced by plant density, diseases, pests, lack of nutrients or lodging. On an average the actual yield is 40-55%, and the water limited yield 60-85% of the potential yield.

The potential yields for Denmark are of the following magnitude:

winter wheat   10-13 t grain/ha (15% water)
summer barley   8-10 t grain/ha (15% water)
summer rape   4- 5 t seed/ha (9% water)
fodder beets   17-24 t DM/ha   (25% top DM)
grass   18-25 t DM/ha   (4-5 cuttings)

REFERENCES

Aslyng, H.C. and Hansen, S. 1982. Water balance and crop production simulation. Model WATCROS for local and regional application. Hydrotechnical Laboratory. The Royal Vet. and Agric. Univ., Copenhagen. 200 pp.
Aslyng, H.C. and Hansen, S. 1985. Radiation, water and nitrogen balance in crop production. Field experiments and simulation models WATCROS and

NITCROS. Hydrotechnical Laboratory. The Royal Vet. and Agric. Univ., Copenhagen. 146 pp.

Jensen, H.E. 1980. Afgrødeproduktion og -kvalitet, lysenergi- of vandudnyttelse, nitrogenbalance og -transport i relation til nitrogen- og vandstatus. Eksperimentelle studier med Lolium perenne (summary in English). Disputats. Hydroteknisk Laboratorium. Den kg. Veterinaer- og Landbohøjskole, København. 373 sider.

Keulen, H. Van and De Milliano, W.A.J. 1984. Potential wheat yields in Zambia - a simulation approach. Agric. Systems 14: 171-192.

Nielsen, N.E. 1980. Forløbet af rodudvikling, naeringsstofoptagelse og stofproduktion hos byg, dyrket på frugtbar moraenelerjord. Medd. 1119, Afd. Pl. Ernaering, Den kgl. Veterinaer- og Landbohøjskole, København. 46 sider.

Discussion Session 2.3

Question from Mr. P.E. Rijtema to Mr. A.L.M. van Wijk

Crop development stage and crop development may differ considerably
when the speed of crop development is strongly dependent on soil and
air temperature in the early stage of growth. This is in particular
true for the development of potatoes, maize and beans. I really have
my doubts whether this system can be applied to years with deviating
temperature in May, as for instance was the case in 1964, when average
May temperature was about 3°C higher than normal, resulting in a growth
explosion of the potato crop. On the other hand high temperatures can
reduce the length of the growing season considerably as is for instance
well known for tulips. A short period of high temperatures early June
may shorten the vegetation period by about 10 days, giving a reduction
in production of about 1500 kg/ha. How do you determine harvesting
time in your simulation because it may differ from year to year? In my
opinion the approach of early crop development and temperature sum as
shown by Mr. Aslyng will be better, since it is dependent on temperature conditions and time instead of time alone.

Answer of Mr. Van Wijk

The LAI-generating procedure incorporated in the presented version of
the SWATR/CROPR model is derived for consumption and starch potatoes
forming the greatest part of potato production in the Netherlands.
Normal practice in growing potatoes is nowadays to keep the crop green
as long as possible by sprinkling irrigation and N-fertilisation. Natural dying of the crop hardly occurs. The end of the growing period is
regulated by spraying. Therefore, in consultation with people of the
Extension Service we fixed the harvest date in the model on 15 September for consumption and on 15 October for starch potatoes.
I agree with your remark on the significance of the temperature in the
early stage of growth. In the computation of the development stage in
the model CROPR there is certainly accounted for the influence of the
temperature. The development stage (D_s) is calculated on basis of the
development rate (d_r) after:

$$D_s = \int_0^t d_r \, dt$$

In the vegetative phase ($D_s < 0.5$): $d_r = 0.0136 \times f(T) \times f(\text{day-length})$;
in the generative phase ($D_s > 0.5$): $d_r = 0.0238 \times f(T)$

Questions from Mr. P.J.T. van Bakel to Mr. A.L.M. van Wijk

1) How does the model know beforehand the harvesting time?

2) Regional groundwater flow (seepage), sprinkling irrigation and surface water manipulation is not incorporated in the model, but don't they influence optimal drainage depth?

Answer of Mr. Van Wijk

For your question on t_{harv}, see the answer to the question put by Mr. Rijtema.

I agree with you. Additional water supply by subsurface irrigation and seepage are already incorporated in the model. We did not use them because in this study we abstracted from specific situations. The objective of the presented study was to compare drainage effects on different soils and different crops. Combination of the model with existing models for regional groundwater flow and models developed for surface water manipulation needs not to be difficult.

Questions from Mr. A. Armstrong to Mr. A.L.M. van Wijk

1) Technical questions relating the soil water models:
   - What is the time step used in the calculations?
   - Is there a continuity between the two models, FLOWEX and SWATRE?

2) You present results for sandy and sandy loam soils. Do you have experience with the use of models in clay soils, and how well does it perform in these soils?

Answer of Mr. Van Wijk

The time step allowable to prevent unstability of simulation computations is determined by stability criteria in which the change of hydraulic conductivity with changing water content and depth increment are the main parameters. Generally the time step varies between 0.01 and 0.05 day.

We apply the models also on clay soils. But we are aware to interpret the results with prudence because the models do not account for cracking. Therefore the Institute for Land and Water Management Research (ICW) started a research on the mechanical behaviour of heavy clay soils to derive relationships between moisture content and volume changes of the soil to adapt existing models to swelling and shrinking soil types.

142

Questions from Mr. S.E. Olesen to Mr. A.L.M. van Wijk

1) Assuming that your physically based model operates with certain
horizontal soil layers, I would like to know if the model is sensitive to the thickness of soil layers?

2) In light of the diurnal fluctuations in the water content of the
surface layer please elaborate a little more on how you handle evaporation from the bare ground in spring.

Answer of Mr. Van Wijk

In the model FLOWEX we apply generally a layer thickness of 10 cm.
However, it may be larger or smaller. Increasing thickness gives a reduction in accuracy. Moreover the length of model time steps allowable
to prevent model unstability depends on layer thickness. Also computing time increases proportionally with an increasing number of layers.
Reduction of soil evaporation due to drying of the top cm's of the
bare soil is realized in the model through a relation between soil
evaporation rate and the soil water pressure head prevailing in the
top layer.

Question from Mr. J. Bouma to Mr. J. Feyen

You mentioned matching factors for the model of 0.001 for the Belgian
site and 0.1 for the Iranian site. Can you explain which aspect is being matched?

Answer of Mr. Feyen

The matching factor ξ, called the flexibility factor by Feddes et al.
(1978), stands for the effect of non-controllable influences on the
field. In our paper we assumed that the experimental conditions at the
Hélécine site were far more optimal - better controlled - than at the
Ghazvin site, and this explains the difference in order to magnitude
allocated to the matching factor.

Answer of Mr. Van Wijk

The parameter ξ in the CROPR-model reflects the degree of curvature of
the non-rectangular hyperbolic crop production curve; a very small value of ξ results in hyperbolas close to their asymptotes. In principle
ξ should be calculated from experimentally determined response curves.
Usually $\xi = 0.01$ is taken.

Questions from Mrs. Venezian-Scarascia to Mr. J. Feyen

1) I would like to ask Mr. Feyen if the calculation of the ETP was the
same for both locations in Belgium and in Iran and if so, was it

143

verified to be equally suitable for both regions with so different climate?

2) You talked about some parameters to use as input in the model. Do you think that there are more parameters equally important that should be considered like plant density, length of biological cycle etc.?

Answer of Mr. Feyen

The reference evapotranspiration ET_{ref} for a grass cover was calculated for both sites with the modified Penman equation as given by Bierman et al. (1981). The calculated values were tested against measured data. The effect of plant density on the LAI-extension is been taken into consideration by the RU variable (see eq. 5). For winter wheat it is assumed that plant density affects LAI-extension only during the initial stage until tillering. From then on it is assumed that differences in plant density are smoothed out by tillering. The beginning and end date of the different growth stages should be generated by the model itself, when the model must be useful for practical application. In my presentation it still is assumed that DB, DLM and DMA are known. Attempts are going on by our team to model the development of the crop using the temperature sum, daylength and the intercepted photosynthetic active radiation as input.

Questions from Mr. J. Bouma to Mr. H.C. Aslyng

How is your root zone capacity defined? What are the pressure head limits assumed? Can you ignore water contributions from the water table?

Answer of Mr. Aslyng

Effective depth of rooting is the depth to which there is at least 0.1 cm root/cm^3 soil. Root zone capacity is the soil capacity for plant available water down to the maximum effective depth of rooting. Instead of pressure head we use fractions of the actual root zone capacity exhausted by the crop as stated in the paper. In general, in Denmark we can ignore contributions from the water table.

Question from Mr. H. Wösten to Mr. H.C. Aslyng

Are you happy with the less detailed model you used as compared to the detailed SWATRE model?

Answer of Mr. Aslyng

The less detailed model requiring relatively few variables and param-

144

eters is better suitable for larger scale advice on irrigation, fertilization, land use, national water balance, etc.

Question from Mr. B. Lesaffre to all speakers

In order to assess the workability, you measure the soil water pressure
head at 5 cm depth. Did you attempt to use other criteria, such as soil
moisture profile or water table depth?

Answer of Mr. Van Wijk

Yes. During the years 1979 and 1980 we measured regularly soil water
pressure heads at depths of 5, 10, 15, 20, 40, 60, 80 and 100 cm and
groundwater table depth in 10 main arable soils. However, the farmers'
appraisal of workability were correlated best to the pressure heads
at 5 cm depth.

Answer of Mr. Feyen

Our model attempts so far have not been orientated to the prediction
of workability. Most of the time field workability does not interfere
with the farming practice of winter wheat, which is sown in the fall
on a not too wet soil.

Answer of Mr. Aslyng

In testing our model we use actual soil moisture profiles in the root
zone obtained from weekly measurements with neutron scattering down to
the needed depth, which can be 1.70 m when the roots are fully developed.

Question from Mr. A. Armstrong to Mr. A.L.M. van Wijk, Mr. J. Feyen and Mr.
H.C. Aslyng

I would like to ask all three speakers on the adequacy of their assumptions about root development. To what extent are their models based on
validated field observations, and do we need more intensive field
studies on root development?

Answer of Mr. Van Wijk

The maximum rooting depth and the time it is reached depend on both
soil profile and crop considered. They are prescribed in SWATR on the
basis of field investigations. Root development over the period between
emergence date and time of reaching maximal rooting depth is described
through a linear function. But we are aware that most crops show a
shallower rooting depth in wet than in dry years. For long time series
we need more information about root development.

Answer of Mr. Feyen

I agree that the assumptions used to describe root development are rather empirical and not well verified by field observations. The calculated root depth for the Hélécine experiment as given in Figure 2 of my paper corresponds roughly with field observations. To come in the future with more reliable procedures we should analyse more field data.

Answer of Mr. Aslyng

We have used previous studies as a basis. In the model testing we control root development by measuring the soil water profile by neutron scattering in the non-irrigated plots. Neutron scattering is also used for testing the water balance.

Question from Mr. G. Spoor to Mr. A.L.M. van Wijk, Mr. J. Feyen and Mr. H.C. Aslyng

Would the speakers comment on whether there is a need for further work analysing root development to improve this part of their models, or do they feel the present coarse assessment methods are adequate, when field variations in root development are taken into account?

Answer of Mr. Van Wijk

See also the answer to Mr. Armstrong's question. I think there is enough material available on root development of the main crops. What we need for modeling purposes is an analysis of the relationship between root development and soil moisture conditions during the growing period to account for differences in root growth in dry and wet years.

Answer of Mr. Feyen

Because water is neither limiting nor in excess as on most of our loamy soils I have the impression that the procedure presented by Rasmussen and Hanks (1978), see also eq. (4) in our paper, gives realistic root development data for winter wheat. The question still not answered is whether the procedure will work under adverse conditions.

Answer of Mr. Aslyng

For soil types, climatic regions and plant species for which the rooting properties or root growth are 'known' further analysis is not required.

Question from Mr. G.A. Oosterbaan to Mr. A.L.M. van Wijk, Mr. J. Feyen and Mr. H.C. Aslyng

For what practical purposes have the models presented been applied so far (for instance advisory work, estimation of water management projects) and what do you expect from practical application in say the next 5 years?

Answer of Mr. Van Wijk

Results of the presented integrated model approach are used by the Governmental Service for Land and Water Use for the evaluation of the economic effects of improvement of the water management in land consolidation schemes. The model will also be applied in a joint land evaluation study in the United Kingdom and the Netherlands in the context of the 1985-1988 CEC land evaluation programme. Other possibilities are the use of the model FLOWEX for time scheduling of field operations on farms as far as depending on workability and trafficability and the use of SWATR-CROPR for irrigation scheduling and water harvesting.

Answer of Mr. Feyen

Up to now we tried to validate the performance of the SWATRE-CROPR model to varying climatic-soil conditions, mainly with regard to the response of winter wheat to supplemental irrigation. Tests have been made for following sites: Hélécine (Belgium), Alexandria (Egypt), Ghazvin (Iran) and Ladhiana (India). With the experience gained and the modifications introduced I do believe that we soon will be able to use the approach as a management tool for scheduling irrigation under conditions of non-limited and limited water resources.

Answer of Mr. Aslyng

Our models have given a much better knowledge on the regional water balance. They can be used for water allocation and for irrigation management. Further for land use evaluation, for deciding upon time and quantity of nitrogen application and to obtain information on nitrogen leaching under different conditions. Modeling is also greatly promoting efficient planning and working in research and experiments.

Session 3
Installation and maintenance of drainage systems

Chairman: B.LESAFFRE
CEMAGREF, Antony Cédex, France

Soil invasion into drain pipes

W.DIERICKX
National Institute of Agricultural Engineering, Merelbeke, Belgium

ABSTRACT

Envelope materials are used to prevent soil particle invasion into drain
pipes. Although the need for envelopes in some soils is not clear, also the
kind of envelope to use is still unsolved. Actually only coconut fibre envelopes are applied in Belgium. Some other alternatives were tried on experimental fields. A practical evaluation about the need for envelopes and the
protective function of the various treatments was done by endoscopy. From
this endoscopic research it follows that the clay content of the soil is not
the only criterion for the need of envelope material. The organic coconut
fibre is sensitive to deterioration. Stabilisation of the trench backfill
by means of soil conditioners is not very effective to prevent soil particle
invasion. Alternatives like tangled synthetic fibres instead of coconut
fibres may also create soil invasion problems. Hence there is an urgent need
to solve the problems about envelope materials.

1. INTRODUCTION

A serious problem in the drainage of agricultural lands is the invasion
of soil into drain pipes. This soil invasion problem mainly depends on the
structural strength of the soil. Cohesionless sandy soils do not have any
structural strength and drain pipes need an envelope, while stable structured cohesive clay soils, which resist the action of flowing water, do not
require envelopes. For intermediate soils the need for envelopes largely
depends on the acting hydraulic gradient and on their structural stability.
Structural stability does not only depend on the mechanical soil composition,
such as clay content or clay/silt ratio, but also on the physical soil conditions, such as the initial water content before wetting, and on the chemical composition of the soil. Envelopes are required when the acting hydraulic gradient exceeds the critical hydraulic gradient whereby the disruptive
forces exerted by the flowing water cause structural destruction.

Apart from cohesionless sandy soils and structural unstable soils it is
not easy to decide upon the use of envelope materials. When the use of an
envelope is decided upon, the most commonly used material in Belgium, the
Netherlands and probably also to a large extent in West-Germany to date is
the organic coconut fibre envelope. From coconut fibre it is known that it
decays in a more or less rapid way, in some cases even within one year
(Meyer and Knops, 1977).

151

Some experimental fields were installed to evaluate the treatment of the trench backfill with soil conditioners in case plain drains, drains wrapped with coconut fibres or with tangled synthetic fibres were used. All these pipes were visually inspected by endoscopy and a comparative evaluation of the soil invasion has been made.

2. EXPERIMENTAL FIELDS

Although the experimental fields were set up to evaluate the effect of trench backfill stabilisation on drainage function, they also served to collect data about the decay of coconut fibre and on soil invasion. A first experimental field was installed in October 1978 on grassland. The soil had a sandy clay loam texture of which the mechanical composition at several depths is given in Table 1.

TABLE 1   Mechanical soil composition of the first experimental field

Size	Depth			
	25 cm	50 cm	75 cm	100 cm
< 2 μm	20.2%	6.0%	10.2%	0 %
2-50 μm	26.2%	29.0%	40.0%	32.4%
>50 μm	53.6%	65.0%	49.8%	67.6%

Since the drains are installed at a mean depth of 75 cm they are located in a loam soil. The drains installed are:

- plain drains
- plain drains and stabilised trench fackfill
- drains wrapped with coconut fibre
- drains wrapped with coconut fibre and stabilised trench backfill

Where the soil was stabilised the trench backfill was treated with a polyacrylamide solution of 0.2%.

The second experimental field was installed in September 1980 on arable land. The soil has a sand texture of which the mechanical composition at several depths is given in Table 2.

TABLE 2   Mechanical soil composition of the second experimental field

Size	Depth		
	30 cm	60 cm	90 cm
< 2 μm	4.7%	3.8%	3.0%
2-50 μm	7.1%	5.5%	5.1%
>50 μm	88.2%	90.7%	91.9%

Drain depth is only 60 cm because of the water level in the main ditches. The following combinations were used:

- plain drains
- plain drains with stabilised trench backfill
- drains wrapped with coconut fibres
- drains wrapped with coconut fibres and stabilised trench backfill
- drains wrapped with tangled synthetic fibres and stabilised trench backfill

The third experimental field was installed in September 1981 also on arable land. The soil has a clay loam texture and the mechanical composition at several depths is given in Table 3.

TABLE 3   Mechanical soil composition of the third experimental field

Size	Depth				
	Top soil	25 cm	50 cm	75 cm	100 cm
< 2 μm	27.5%	27.2%	28.7%	24.1%	22.3%
2-50 μm	29.6%	31.4%	32.9%	25.0%	27.2%
>50 μm	42.9%	41.4%	38.4%	50.9%	50.5%

Since the drains were installed at a depth of only 70 cm because of the water level in the main ditch, they are in a sandy clay loam texture. The drains installed were:

- drains wrapped with coconut fibres
- drains wrapped with coconut fibres and stabilised trench backfill

The fourth experimental field was on a clay soil, of which no mechanical analysis is available. Drain pipes wrapped with coconut fibres, with a mix-ture of 50% of coconut fibres and 50% of synthetic fibres (on weight basis)

153

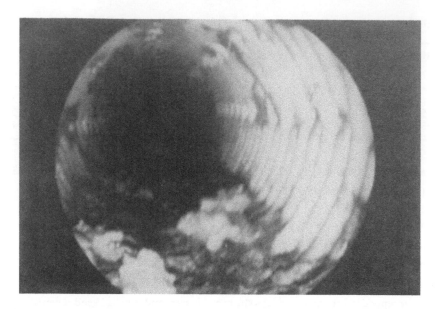

Fig. 1 Endoscopic view of the inside of a drain pipe wrapped with coconut fibre, 6 years after installation in a loam soil

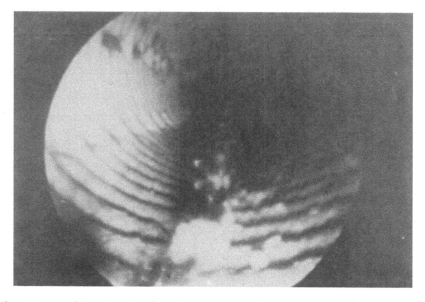

Fig. 2 Inside of a drain pipe wrapped with coconut fibre and covered with stabilised trench backfill, 6 years after installation in a loam soil

and with tangled synthetic fibres only, were installed in October 1979.

3. RESULTS

To get information about the need for envelope materials in different soil types, to find out the protective function of the used envelope materials and to learn the effect of trench backfill stabilisation, an investigation by means of endoscopy was carried out in September and October 1984 on the experimental fields described above.

In the first experimental field soil invasion was observed in all drainage types. From the observations it could not be concluded that protection of drain pipes with an envelope of coconut fibre was better than a plain drain. Within a period of 6 years the coconut fibre was completely decayed and vanished completely. Also the treatments of the trench backfill with soil conditioners did not result in an effective protection against soil invasion. Some results are given in Figure 1 which shows the inside of a drain wrapped with coconut fibre. Figure 2 depicts the inside of a drain wrapped with coconut fibre and covered with a stabilised trench backfill, and Figure 3 shows the inside of a plain drain covered with a stabilised trench backfill.

In the second experimental field where the drains were installed in a sandy loam soil and 4 years old, the coconut fibre envelope was not yet decayed completely and in most cases it still fulfilled its function, as can be seen from Figure 4 in which the inside of a drain wrapped with coconut fibre and covered with stabilised trench backfill is shown. From Figure 5 which depicts the inside of a plain drain covered with stabilised trench backfill, it can be seen that the effect of soil stabilisation does not prevent soil invasion into the drain pipe. The most effective material for that purpose seems to be the synthetic fibre envelope, as can be seen from Figure 6 where the inside of a drain wrapped with synthetic fibres and covered with stabilised trench backfill is shown.

In the third experimental field the drains wrapped with coconut fibre were installed in a clay soil. Three years after installation the coconut fibre envelope was completely decayed and the degree of siltation was the same for drains wrapped with coconut fibre only (Figure 7) and for drains wrapped with coconut fibre and covered with stabilised trench backfill (Figure 8).

155

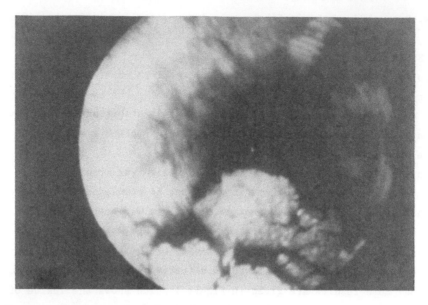

Fig. 3   Inside of a plain drain covered with stabilised trench backfill, 6 years after installation in a loam soil

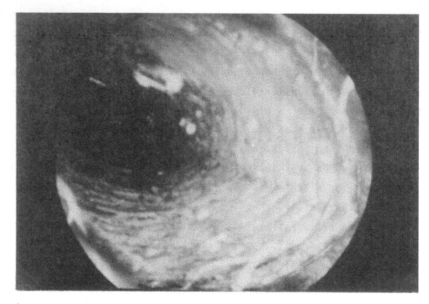

Fig. 4   Inside of a drain pipe wrapped with coconut fibre, 4 years after installation in a sandy soil

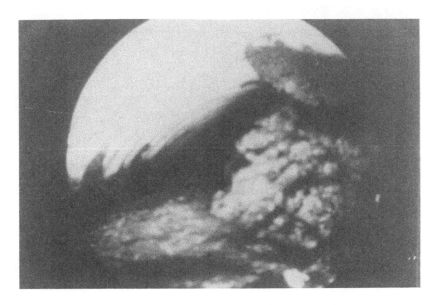

Fig. 5   Inside of a plain drain covered with stabilised trench backfill, 4 years after installation in a sandy soil

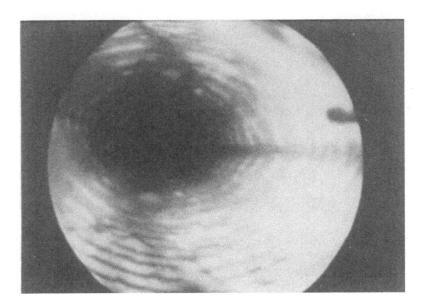

Fig. 6   Inside of a drain wrapped with tangled synthetic fibres and
covered with stabilised trench backfill, 4 years after installation in
a sandy soil

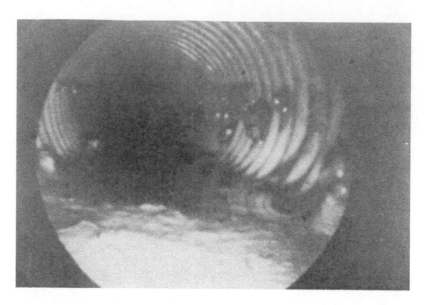

Fig. 7   Inside of a drain wrapped with coconut fibre, 3 years after installation in a sandy clay loam soil

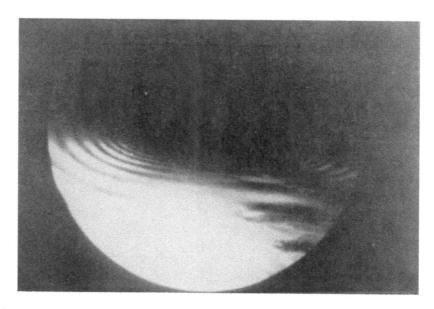

Fig. 8   Inside of a drain wrapped with coconut fibre and covered with
stabilised trench backfill, 3 years after installation in a sandy loam
soil

Fig. 9   Drain pipe wrapped with coconut fibre, dug up 5 years after installation in a clay soil

Fig. 10   Drain pipe wrapped with a mixture of coconut fibre and synthetic fibre, dug up 5 years after installation in a clay soil

In the clay field that was installed in October 1979, the coconut fibre envelope was completely decayed 5 years after installation and the pipes were invaded with soil particles reducing the water transporting capacity of the drain pipes (Figure 9). Although less, the amount of invaded soil material was still important in the case of an envelope consisting of the mixture of 50% synthetic and 50% coconut fibre (Figure 10). Also here no traces of coconut fibre were found in the envelope after 5 years. Due to the de-

Fig. 11   Drain pipe wrapped with tangled synthetic fibre, dug up 5
years after installation in a clay soil

composing of the coconut fibre the pores of the envelope became coarser, resulting in a still considerable soil particle invasion. Less soil material,
although still too much, was found in the pipe wrapped with a complete
synthetic fibre envelope as shows Figure 11, probably because the used envelope material was too coarse.

4. CONCLUSIONS

The results from field experiments clearly illustrate that:

- Stabilisation of the trench backfill does not prevent siltation of drain
  pipes.
- Organic envelopes and especially coconut fibre are sensitive to decay
  sooner or later, hence siltation hazard of drain pipes still exists.
- Tangled synthetic fibre envelopes seem to function well in light textured
  soil. In fine textured soils a large amount of fines are found in the pipe
  and as a consequence they were not effectively protected from siltation.
- When coconut fibre is used in a mixture with synthetic fibre it also
  decays. The remaining synthetic fibre envelope may become too coarse as an
  envelope to prevent soil particle invasion.

Although in the West-European countries preference is given to voluminous
envelope materials of organic origin, in the USA and Canada commercially
available woven and non-woven thin fibre fabrics are used and seem to perform well in light textured soils (Knops and Dierickx, 1979).

160

Organic envelope materials decay so that the siltation hazard of drain pipes still remains. On the other hand we learned that there is evidence that thin fibre fabrics are liable to clogging. The results discussed here and the practical experience of the past clearly show that little is known about the need for envelopes and about the most suitable envelopes to be used. Therefore there is an urgent need to solve the problems of the choice of envelope materials. To date, in many cases tangled synthetic fibre envelope is proposed, but will that be really the solution?

REFERENCES

Knops, J.A.C. and Dierickx, W. 1979. Drainage materials. In 'Proceedings of the International Drainage Workshop' (Ed. J. Wesseling). ILRI-publication 25: 14-38.

Meijer, H.J. and Knops, J.A.C. 1977. Veldonderzoek naar de bestendigheid van cocosvezels als afdek- of omhullingsmateriaal voor draineerbuizen. Cultuurt. Tijdschr. 16,6: 261-265.

Common lines in research on drainage envelopes in France and in the Netherlands

L.C.P.M.STUYT
Institute for Land & Water Management Research (ICW), Wageningen, Netherlands
T.CESTRE
CEMAGREF, Antony Cédex, France

ABSTRACT

The close link between France and the Netherlands regarding the research efforts dealing with drain clogging problems is described.* Despite the fact that pedological conditions and drainage installation practices are widely different, research efforts are complementary and of mutual interest.

After a brief description of some facts and figures about subsurface drainage in both countries, the clogging problem is discussed. Some countermeasures, e.g. the use of envelope materials, are mentioned. In both countries, pilot projects are underway in widely different soil types where the utility of various types of envelopes is investigated. In addition, laboratory research is carried out.

Current research projects are intended to serve the interests of three involved parties: design engineers, contractors and manufacturers of drainage materials. Gradually it has become clear that the study of mineral clogging, the dominant clogging type in both countries, is formerly often underestimated and complex and that the solution requires sophisticated research techniques and equipment.

Research at CEMAGREF and ICW currently focusses at increasing the knowledge of processes occurring at the soil/envelope interface, at assessing the validity of laboratory permeameter flow tests, and at studying the effect of various constructional aspects such as grade and perforation of pipes. The progress reached in these fields until now, and plans for future work are briefly outlined.

1. INTRODUCTION

According to a survey by the French National Drainage Contractors Association (SNED) in 1983 about five million hectares out of a total of 29.5 million hectares of agricultural land in France are incorporated in land amelioration plans. Nowadays over 1.5 million hectares are drained by subsurface drainage systems (5.3%). During the last decade, the installation rate of these systems has grown, and equals ±125 000 hectares annually.

*Part of the research described in this paper was presented jointly by the authors at the 1984 study days of the Syndicat National des Entreprises de Drainage (SNED) in France (Cestre and Stuyt, 1985; Stuyt, 1984)

In the Netherlands, about 900 000 hectares of agricultural lands are
drained by means of pipes, which equals about 45% of the total area of agricultural lands (2.1 million hectares).

Many subsurface drainage systems are prone to clogging, in the ultimate
case leading to system malfunctioning. Clogging by plant roots, by ochre
formation (a microbiological process) and clogging by dissolved salts (e.g.
in coastal regions with seepage) occur occasionally. The most frequently
occurring clogging type is mechanical or internal mineral clogging, i.e.
the washing of soil particles and aggregates into the drains.

Two types of mineral clogging must be distinguished: primary and secondary. Primary mineral clogging occurs immediately upon installation, that is,
during the first discharge wave. Secondary mineral clogging develops gradually during the functioning of the drainage system, and in fact is everlast-
ing.

Primary clogging risks are mainly connected to the condition of the soil
during installation (i.e. water content), the method of installation
(trencher or trenchless) and the meteorological circumstances during installation. Although installation under poor conditions should be avoided the
involved parties often state that this is impossible because of organization-
al and/or economic reasons. When subsequent secondary clogging is not likely
to occur (a matter of experience) organic envelopes like peat- or coconut
fibres, subject to microbiological decomposition, may be used.

Secondary clogging risks are relatively independent of installation conditions, and are the most serious threat for the long-term functioning of
the system. They can be tackled only by long-lasting means. Therefore, envelopes, consisting of synthetic fibres or filaments ('geotextiles') are
preferred in this case. Especially during the past few years, the use of
geotextiles as envelopes in agricultural drainage has increased, and it is
likely that their use will increase even more rapidly in the coming years.
As an alternative, in France, soil stabilizing agents (polymers) are proposed, but their application is still in an experimental stage.

In the Netherlands, the majority of the soils to be drained is unstable
and prone to mineral clogging. As a consequence, over 80% of all drains
need an envelope. In France, this percentage varies between 3.7 and 5.8
(Figure 1) but is regionally bound: it can be as high as 50% in coastal
plains like les Wateringues, where the same soil types are found as in the
Belgian and Dutch coastal regions.

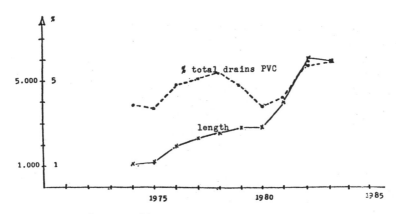

Fig. 1   Wrapped* drains** installed in France in the past decade
(source: annual statistics of the Syndicat National des Entreprises de
Drainage, SNED)
*organic and synthetic materials   **∅ 44/50 and 58/65 mm

2. DEVELOPMENTS IN RESEARCH

The Netherlands, with its relatively great need for wrapped drains, has
a longer research 'tradition' in this field than France. As early as the
1950s, pilot projects were established in clogging-prone soil types. Simultaneously, filter materials were tested in laboratories. Together with the
introduction of smooth plastic PVC pipes, soon followed by corrugated PVC
pipes, new filter materials were needed and proposed. In the beginning of
the 1970s, coconut fibre was introduced as envelope material. Very soon
these envelopes were generally accepted and widely used and today the majority of Dutch envelopes consists of coconut fibres. At the end of the same
decade, doubts arose regarding the lifetime and sand-tightness of cocos-
envelopes. The Dutch Government Service for Land and Water Use ('Landinrich-
tingsdienst') made a considerable number of dig-ups, but concluded that
cocos was a reliable envelope material.

Around 1977, the French established pilot projects in the coastal plains
(les Wateringues du Nord and le Pas du Calais) in the extreme northwest of
the country where alluvial calcareous soils and pseudogleys are common.
Points of resemblance between these areas, and the areas where, in the
Netherlands, pilot plots were installed earlier, are:

- the occurrence of (very) fine sandy soil layers (particle size ranging
  from 50 to 200 μm) having particles small enough to wash into the drain
  and yet large enough to settle into it;

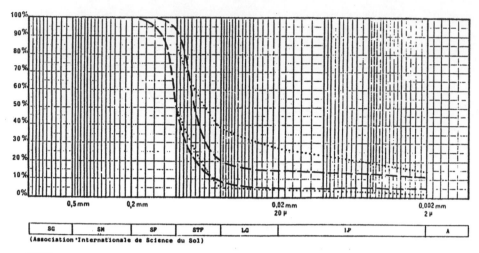

Champlan Sand (used in laboratory tests CEMAGREF)

Wateringues Sand

Fig. 2   Textural composition of soils prone to mineral clogging in France

- very small pipe grades (1 to 2°/oo) or even no grade. Despite the application of contemporary techniques (laser grade control, control of the orientation of the tine of trenchless machines etc.) the risk of increased sedimentation due to erroneous grades still is present. Recent investigations in the Netherlands and in France indicated that this is a serious problem.

Between 1970 and 1975, the first PVC drains laid in the coastal plains in France were wrapped with organic fibres (peat, cocos and straw). Mineral clogging mechanisms were studied here micro-morphologically (Sole-Benet, 1979). Like the Dutch, the French supported their field investigations by laboratory tests, using soil types very similar to those in the pilot plots as regards composition (Champlan sand, cf. Figure 2).

Apart from 'traditional' materials like coconut fibres, the French have installed some synthetic materials as well. Besides filtering data, life expectance data of envelopes were obtained. Simultaneously, Knops et al. (1979) published a detailed and comprehensive set of recommendations for envelope application in Dutch soil types. These recommendations, however, were only partly followed in practice. The recommendations just mentioned are about to be replaced by new ones, reflecting the present state-of-the art.

In France, the last pilot plot in an unstable, fine-sandy soil was installed in 1983, because it was felt necessary to test new products that had been introduced by them. In some French pilot plots, drain discharge and groundwater level data were recorded.

At the end of the 1970s and the beginning of the 1980s, laboratory testing gained increasing attention in the Netherlands. The Dutch Polder Development Authority (RIJP) investigated the applicability of new synthetic products for use in the newly reclaimed IJssellake area (Zuidema and Scholten, 1979) and at the Institute for Land and Water Management Research (ICW) permeameter flow testing techniques were gradually improved (Eskes, 1977; Seijger, 1978; Stuyt, 1982, 1983, 1984).

3. CURRENT RESEARCH

3.1. Aim and scope

Current research activities in both countries serve the interests of three involved parties, namely:

- designers of drainage systems, who need more reliable prediction methods as regards clogging risk, recommendations for selection of envelopes for different soil types and – mainly in France – data indicating the possibilities of 'auto-cleaning' of pipes by using larger grades in non-sandy soils where no envelopes are used;
- contractors, who seek a better characterisation of the envelope materials currently available on the market, a better control of envelope characteristics and more information on the applicability of envelopes in various soil types. These aims, if fulfilled, should lead to a better and more regular quality of envelopes on the market, and in the long term to the normalization of simple laboratory tests for assessing envelope characteristics rapidly and unequivocally. In the Netherlands, most envelope types must comply with criteria established by the Dutch foundation KOMO ('Kwaliteitsverklaringen-Organisatie voor Materialen en Onderdelen voor de bouw'). KOMO working groups may be formed on occasion if market trends require this. Currently, a KOMO working group establishes requirements for voluminous envelopes consisting of polypropylene fibres, wrapped at random around the pipe; a preliminary set of requirements including a comprehensive yet simple test to screen envelopes for these requirements may be expected by the autumn of 1985. In the same way in France; an AFNOR (Association Française de Normalisation) working group is setting

up experimental standards and is testing a draft schedule of specifications in 'les Wateringues';
- manufacturers of pipes and envelopes, and wrapping companies, who want
  to improve the applicability of their products and to market materials
  having well-known characteristics, confirmed by official testing data
  and certificates. At the Institute for Land and Water Management Research
  (ICW) current research is supported financially by eleven contracting
  parties from Europe and North America; in France it is supported by French
  geotextile manufacturers. These projects facilitate regular contacts between research workers and manufacturers which has proved to be highly
  beneficial.

3.2. Regular contacts

In the Netherlands, new developments spread easily due to the small
size of the country and the many contacts between design engineers. Each of
the eleven provinces has its own research unit and drainage engineers meet
at a regular basis.

In France, drainage develops very rapidly in many regions. However, not
in all regions design engineers and technicians of the drainage division
of CEMAGREF are available. Therefore, since 1980, the field work (data
collecting on existing drainage systems and field experiments) is allocated
in so-called 'reference areas' (Favrot et al., 1980). In doing so, a sound
scientific basis can be assured, and a considerable number of data regarding the various soils concerned is obtained. Additionally, the execution of
local field tests is made possible.

In both countries, regular feed-back mechanisms exist that allow for
the incorporation of past experience with drainage into future design. In
France, regular contacts exist between research centres, members of the
SNED, pipe and envelope manufacturers and consulting engineers, who critical-
ly examine and evaluate the functioning of the systems they have installed
and try to improve design and installation methods. Their experience leads
to a better understanding of the possible causes of system malfunctioning
at national scale and offers the possibility to determine research priorities at CEMAGREF and INRA (Institut National de la Recharche Agronomique).
More recently, the National Experiment and Demonstration Network (RNED) has
been created, gathering all the bodies involved in drainage (including ad-
ministration and farmer boards). In the Netherlands, the Dutch Governmental
Service for Land and Water Use organizes meetings between its engineers,

technicians and scientific researchers, involved in drainage, twice annually. This 'Drainage Contact Group' exchanges recent trends, experiences and developments. Researchers of the Institute for Land and Water Management Research (ICW), the Agricultural University Wageningen (LH), the Dutch Polder Development Authority (RIJP), together with staff-members of Engineering Bureaus have regular meetings in the 'Drainage Study Group', where scientific developments in the Netherlands and abroad are discussed. Belgium is represented in the Study Group as well.

3.3. Research topics

Over the years it has become clear that, in order to understand the causes of drainage system malfunctioning due to pipe and/or envelope clogging, field investigations must be detailed and executed carefully. They are therefore expensive and time-consuming.

Progress in research in this field has reached a level, as is acknowledged in both countries, that simple dig-ups, followed by visual inspection only generally do not render much new information. Better observational techniques, however, are being developed. In France, micromorphological techniques on undisturbed soil/envelope/drain sediment samples are used. In the Netherlands, non-disruptive visual inspection of drain sediments in situ ('drain-o-scope') is applied, as well as electronic analysis of the textural composition of samples of the soil, drain sediments and suspended soil material, by means of sophisticated particle size analyzing equipment.

Current research at ICW and CEMAGREF has in common that it aims at a better understanding of the mechanisms of particle migration at and in the vicinity of the soil/envelope interface. At both research centres, laboratory permeameter flow tests are performed, with many aspects in common, and in fact only different as regards the soil types used.

At CEMAGREF, aggregate stability is investigated following earlier work by Dierickx (Belgium) and Willardson and Samani (USA). Permeameter flow tests are performed with air-dried and sieved (2 mm) aggregates or sand particles (Bluhm, 1985). At ICW, both disturbed and undisturbed soil samples are taken at drain depth, and are used in the permeameter flow tests. In doing so, the influence of soil structure is incorporated in the testing programme. Because soil samples are installed in the permeameters the very same day, having a moisture content existing in the field at drain depth, the aspect of installation conditions are incorporated to a certain extent too. The flow direction in permeameter tests at CEMAGREF is upward (the most

169

destabilizing direction); at ICW, five directions (upward, downward, horizontal and oblique) can be applied.

During permeameter tests, soil/envelope combinations are subjected to increasing hydraulic gradients and recordings are made of:

- the permeability changes at different heights in the soil core, using piezometer readings;
- the critical hydraulic gradient, at which particles start to move through the envelope material and/or the soil permeability rapidly decreases;
- the permeability and filtering action of envelopes in the course of time.

In the experimental setup used by ICW, the following additional observations are made:

- electronic analysis of the textural composition of suspended sediments during the two-week flow tests;
- after termination of the flow test: electronic analysis of micro soil samples from the soil core to detect the movement of soil particles in the soil core due to the hydraulic forces applied;
- after termination of the flow test: electronic analysis of the textural composition of the soil material entrapped in the envelope material.

The last two items mentioned will be incorporated into the research at CEMAGREF too in the near future.

The results of the permeameter flow tests, in fact performing as soil stability tests, are of comparative nature only.

Studies are carried out with different soil materials of varying texture (in the Netherlands, four sites have been selected) originating from experimental fields or pilot plots. In doing so, it must be possible to calibrate laboratory permeameter tests to field conditions. In order to facilitate this calibration in the ICW programme, soil/envelope/drain samples will be taken in pilot plots, conserved carefully by plastering the sample with gypsum, brought to the laboratory, and studied carefully for hydraulic conductivity, particle size composition, etc.

These research efforts by the two countries must lead to:

- a definition of the relation between soil characteristics and mineral clogging risk;
- a definition of risk 'thresholds' following from field sample observations;
- a definition of preferable soil/envelope combinations.

170

At CEMAGREF, special attention is given to water flow in the vicinity of the pipe (Tiligadas, 1984). This study is based upon earlier work by Dierickx (1980), Nieuwenhuis and Wesseling (1979) and Wesseling and Homma (1967). It aims at precising the distribution of equipotential lines around naked or wrapped drains, for different wrapping materials, hydraulic gradients and water flow boundary conditions. Monitoring flow rates and hydraulic gradients in large soil cubes, drained at their centres, and supplied with water from the bottom, the top and laterally should increase our understanding of water flow around the drain, and consequently, of mechanisms of mineral clogging: destabilisation, particle transport in the soil and towards the perforations, in relation to permeameter observations (critical hydraulic gradient, hydraulic conductivity changes, etc.).

At CEMAGREF, the additional head loss due to the convergence of streamlines towards drain pipes (radial resistance) was studied for various perforation grades, patterns and diameters, using non-clogging glass beads (∅ 1.4 mm), thus creating a uniform porous medium which is reproducible and has known values of porosity and hydraulic conductivity (Lennoz-Gratin, 1984). The results of this study may eventually be used to revise the French norm NF (AFNOR; Association Française de Normalisation) for drain perforations (CEMAGREF, 1985).

In the Netherlands where, just like in Belgium and Germany, 'voluminous, coarse' envelopes (that is envelopes having effective pore diameters ranging from 200 μm to 800 μm, and a thickness exceeding 3 mm) are the most frequently used types, wetting problems were virtually absent. Consequently, no research into this aspect was done. With the increasing application of geotextiles, however, wetting problems gradually attract more attention, and are incorporated in ICW's current research programme. In France, with its longer history of geotextile use, wetting problems, noted regularly by contractors and other field workers, have given rise to investigations at CEMAGREF, partly following proposals by manufacturers (Cestre, 1985). For this project, CEMAGREF has compared several measuring devices available on the market, and has modified a device in such a manner that it allows for measuring rapidly:

- the hydraulic head, required for an envelope material to start transmitting water, and
- the area of the envelope material, effectively involved in the water flow domain.

These experiments should ultimately lead to an experimental AFNOR-norm so that the contractors can dispose of a measuring device enabling them to assess the wetting characteristics of drain envelopes by themselves. Cooperation between ICW and CEMAGREF is expected in this field.

The influence of the grade of corrugated drains on the sedimentation/ suspension balance of particles washed into the pipe is investigated at CEMAGREF (Ben Tekaya, 1984). A model was constructed consisting of a 16 m long corrugated pipe, Ø 44/50 mm, the upper half removed, with adjustable grade (1 to 20°/oo), and flow rate. Recorded were the development of the deposit of fine sand (e.g. the formation of sand wrinkles and their movement ahead), the solid and liquid flow rates at the downstream end of the pipe, and the flow rate distributions.

REFERENCES

Ben Tekaya, N. 1984. Drain annelé en P.V.C.: étude du transport solide en écoulement à surface libre. ENGREF-CEMAGREF.

Bluhm, H. 1985. Colmatage minéral des drains. Mise au point d'un test de stabilité en perméamètre. CEMAGREF (in press).

Cestre, T., Chossat, J.C., Lesaffre, B. and Tiercelin, J.R. 1983. Le colmatage des drains agricoles. Etat des connaissances en France en 1983. Informations techniques no. 51. Revue Drainage.

Cestre, T. 1985. Mouillabilité des géotextiles utilisés en enrobage de drains agricoles (in press).

Cestre, T. 1985. Drainage des sols sensibles au colmatage minéral. Revue Drainage no. 28/29.

Dierickx, W. 1980. Electrolytic analogue study of the effect of openings and surrounds of various permeabilities on the performance of field drainage pipes. Comm. of the Nat. Inst. for Agric. Eng. (Meded. Rijksstation voor Landbouwtechniek) 77, Merelbeke.

Eskes, B.T.F. 1977. Laboratoriumonderzoek naar de poriënverdeling, de filtrerende en de hydrologische eigenschappen van synthetische omhullingsmaterialen voor drainbuizen. Internal report, International Institute for Land Reclamation and Improvement (ILRI) and Institute for Land and Water Management Research (ICW), Wageningen.

Favrot, J.C., Horemans, P. and Lesaffre, B. 1982. Pratique des secteurs de références. Revue Génie Rural.

Knops, J.A.C., Zuidema, F.C., Van Someren, C.L. and Scholten, J. 1979. Guidelines for the selection of envelope materials for subsurface drains. Proc. Int. Drain. Workshop (ed. J. Wesseling). ILRI-Publication 25, Wageningen. 731 pp.

Lennoz-Gratin, C. 1984. Drains en P.V.C. annelés: influence de la taille des perforations sur les écoulements et l'interface sol-drains.

Michel, R. 1985. La marque NF drains agricoles. CEMAGREF, Informations Techniques no. 54. Revue Génie Rural.

Nieuwenhuis, G.J.A. and Wesseling, J. 1979. Effect of perforation and filter material on entrance resistance and effective diameter of plastic drain pipes. Agric. Water Managem. 2: 1-9.

Seijger, L.G. 1978. Laboratoriumonderzoek van omhullingsmaterialen voor drainbuizen. Nota 1088 ICW, Wageningen.

Sole-Benet, A. 1979. Contribution à l'étude du colmatage minéral des drains. CEMAGREF, mémoire no. 13.

Stuyt, L.C.P.M. 1982. Stochastische simulatie van de bepaling van een pF-curve van grofkorrelige materialen. Nota 1336 ICW, Wageningen.

Stuyt, L.C.P.M. 1983. Laboratoriumonderzoek aan drainage-omhullingsmaterialen: een interim-rapportage. Nota 1436 ICW, Wageningen.

Stuyt, L.C.P.M. 1983. Drainage envelope research in the Netherlands. Report 8 ICW, Wageningen. Proc. 2nd Int. Drain. Workshop, Washington DC. CPTA, Carmel, IN: 106-123.

Stuyt, L.C.P.M. 1985. Recherches de laboratoire concernant les sables fins sensibles au colmatage des tuyaux et des matériaux filtrants. Revue Drainage no. 28/29.

Tiligadas, E. 1984. Hydraulique au voisinage du drain agricole. Thèse de docteur-ingénieur, Université Paris 6.

Wesseling, J. and Homma, F. 1967. Entrance resistance of plastic drain tubes. Techn. Bull. 51 ICW, Wageningen. Neth. J. Agric. Sci. 15: 170-182.

Zuidema, F.C. and Scholten, J. 1979. Model tests on drainage materials. Proc. Int. Drain. Workshop (ed. J. Wesseling). ILRI-Publication 25, Wageningen. 731 pp.

Use of hydrographs to survey subsurface drainage: Networks ageing and hydraulic operating

B.LESAFFRE & R.MOREL
CEMAGREF, Antony Cédex, France

ABSTRACT

As a result of new concerns about subsurface drainage (networks ageing, hydraulic and hydrological operating), several methods have been developed to analyse continuously recorded field drainage experimental data.

This paper presents these methods and their results, illustrated by surveys of some field drainage experiments located on various shallow temporarily waterlogged soils, three of them having been monitored during eight years.

Annual and interannual hydrological performance of drainage is studied by double-mass method. It highlights drainage seasons, enables a comparison between different kinds of layouts on a given soil, shows the contrasts in hydraulic behaviour between different soils and illustrates the drainage network performance particularly under tilled conditions.

A statistical study of both maximum flow rates with different return periods and durations of given flow rates is conducted. This analysis of high flow rates is of practical interest for the choice of drainage design rates with different allowable surcharge duration.

1. INTRODUCTION

The acreage of French waterlogged lands is assessed to be about 10 million hectares i.e. one third of the total agricultural land. Half a million hectares is permanently waterlogged, due to shallow groundwater tables. The rest, therefore most of French soils, are temporarily waterlogged, because of either flow from outside (runoff, springs) or storage of rainfall inside a fairly permeable and often shallow horizon lying above an impervious layer (temporary perched water tables) (Devillers and Guyon, 1978). Heavy soils can be considered as a sub-category of this type of soils (Bouzigues et al., 1981).

Due to the rise of the annual acreage of subsurface drainage (130 000 ha in 1983), drainage works are more and more in difficult soils, with ever stronger economic constraints. Therefore hydrograph data have been collected for a large number of experimental networks; theoretical models and data analysis methods have been improved.

Data from several networks have been recorded with small time steps (one hour) and over long periods (up to a decade), thus enabling statistical analysis. Three field experiments (Arrou, La Bouzule, Longnes, Figure 1) installed under more or less oceanic climate and in different temporarily

TABLE 1 Some characteristics of the experiment locations

Heading/location	Arrou (Eure et Loir)	Longnes (Yvelines)	La Bouzule (Meurthe et Moselle)
- Available measuring periods (a period begins in September and finishes in August; rainfall and discharge hydrographs are recorded on magnetic files; record time step: one hour)	8 periods (1974/75 till 1981/82)	8 periods (1971/52 till 1978/79)	8 periods (1974/75 till 1981/82)
- Climate	oceanic	altered oceanic	with a semi-continental tendency
. mean annual rainfall	620 mm (Chateaudun 1931/60)	490 mm (Dreux 1931/60)	710 mm (Nancy-Tomblaine 1931/60)
. 3 days rainfall with return period one year (except for summer)	11 mm/day	9 mm/day	14 mm/day
- Parent material	Plateau loam on flint clay	Plateau loam on Romainville's green clay	clay marl with sinemurian hippopodium · clay marl blanketed by silt
- Soil	hydromorphic leached soil on nearly impervious flint clay 60 to 120 cm deep	leached on nearly impervious clay, 90 to 110 cm deep, with a shallow stonebed	brown pelosol · leached brown soil with a pseudo-gleyed layer
. texture	clayey silt (50–130 cm)	sandy clayey silt (20–100 cm)	clay (about 55% of mainly swelling clay) · clayey silt (40–100 cm)
. permeability (in situ)	0.3 m/day	1.4 m/day	<0.1 m/day · <0.2 m/day
. effective porosity (in situ)	1.4%	3%	– · –
- Drainage lay-out			
. plot size	8 x 2 ha	1.89 and 2.03 ha	1.85 ha · 2.83 ha
. drain spacing	10 – 15 – 20 m	16 m	8 m · 12 m
. technical installation	non-enveloped PVC or clay pipes installed at 80 cm depth by a trench machine in 1972, with either a natural or a gravel backfill	non-enveloped clay pipes installed at 90 cm depth by a trench machine in 1960	non-enveloped PVC pipes installed at 90 cm depth by a trench machine in 1969
. drainage design rate	$1.0 \ l \cdot s^{-1} \cdot ha^{-1}$	$1.0 \ l \cdot s^{-1} \cdot ha^{-1}$	$1.0 \ l \cdot s^{-1} \cdot ha^{-1}$
. average slope	0.5%	.4%	6% · 1%

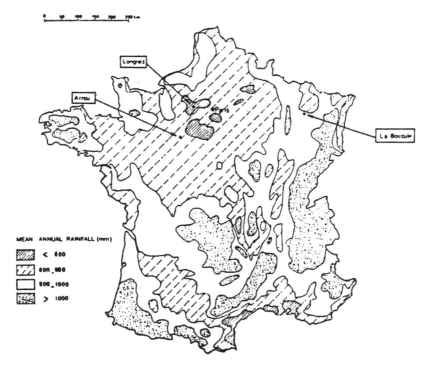

Fig. 1   Field experiments location

waterlogged soils are considered here.

This paper first presents the experimental layouts and monitoring, and some of the present knowledge about drainage operating in such an environment. It will them deal with:

- description of hydrological performance through the double-mass method, with emphasis on network ageing characterization and its possible causes;
- description of hydraulic performance through statistical analysis of peak flows, with a discussion on the choice of the design drainage rate.

Indications on further research will ultimately be given.

2. EXPERIMENTAL LAYOUTS

Environments and experimental layouts are briefly described in Table 1 (Guyon, 1966, 1983; Florentin, 1982; Laurent and Lesaffre, 1983). In each case, a perched water table forms in winter in a shallow horizon (60 to 120 cm thick), poorly to fairly permeable (ranging from 0.2 to 1.4 m/day). At La Bouzule, the heavy swelling clay is drained without a secondary treatment (moling or subsoiling).

177

The monitoring system includes a rainfall gauge and a flow-rate room equipped with V-notch weirs and groundwater level recorders. Measurements of the groundwater table (discrete and continuous at Arrou), of the soil water content and tension (at Arrou and La Bouzule) have been carried out as well; water samples have been collected automatically at a rate proportional to the outflow.

Data collecting and recording systems have quickly improved recently: from early 1985 on, new gauges measure water level through ultrasonics and record it on an electronic micro-chip. But whatever their origin, all data used in the present paper are stored on magnetic floppy disks.

3. REVIEW OF HYDRAULIC OPERATING OF SUBSURFACE DRAINAGE IN TEMPORARILY WATER-
   LOGGED SOILS

Some authors noticed, either a long time ago already (Russel, 1934) or more recently (Van Hoorn, 1973; Trafford, 1973; Al Soufi and Rycroft, 1975; Alessandrello et al., 1976; Herve, 1980; Florentin, 1982) a large variation of flow rates out of drainage networks installed in waterlogged soils (Lesaffre and Laurent, 1983). A typical winter discharge hydrograph comprises two stages (Figure 2):

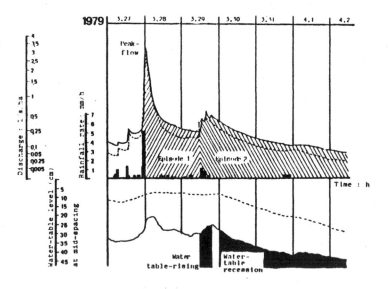

Fig. 2   Arrou's drainage hydrographs
plot 1 (S = 10 m).———   plot 4 (S = 20 m) -------

178

- the peak-flow stage, during which discharges increase and decrease very
  quickly as a response to rainfall; the discharge values are high and
  their durations are short;
- the tail recession stage, which corresponds to the falling water table,
  in periods without rainfall; the discharge values are low and their durations are long.

During the latter stage, water table and flow rates constantly decrease,
according to well-known unsteady state formulas, based on laws of hydrau-
lics of groundwater flow (Guyon, 1961, 1966; Van Schilfgaarde, 1963, 1965)
that are confirmed through field experiments. Drain spacing is usually calculated from the assessment of time required for a proper trafficability.
The basic formula is Guyon-Van Schilfgaarde's, valid for homogeneous and
isotropic soils, with drain pipes located on or in the impervious layer:

$$t = \frac{\alpha}{2} S^2 \frac{f}{K} \left(\frac{1}{h(t)} - \frac{1}{h_o} \right) \tag{1}$$

where S = drain spacing

  K = hydraulic conductivity

  f = effective porosity

  h_o = initial water table level above the impervious layer at mid-
      spacing (about the basis of the tilled layer)

  $h(t)$ = final water table level above the impervious layer at mid-
      spacing (usually derived 45 cm in France)

  t = water table drawdown lag (from h_o to $h(t)$), often taken as 1
      day

  α = first water table shape coefficient (usually $\alpha/2 = 2/9$)

Guyon (1980) extended eq. (1) to anisotropic and vertically heterogeneous soils, introducing the concept of equivalent horizontal hydraulic conductivity $\tilde{K}_h(h)$ that can be measured in situ (Guyon and Wolsack, 1978):

$$\tilde{K}_h(h) = \frac{2}{h^2} \int_o^h K_h(z)(h-z)dz \tag{2}$$

$K_h(z)$ is the horizontal component of the local hydraulic conductivity, only
depending on the elevation z, measured from the impervious layer.

Under some assumptions, confirmed through Arrou's field experiment
(Guyon, 1983), eq. (1) becomes:

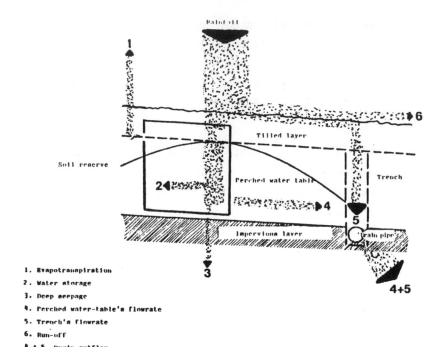

Fig. 3   Temporarily waterlogged soil, subsurface drainage (sketch)

$$h(t) = \frac{h_o}{(1 + \beta t)^n} \tag{3}$$

where $\beta = \dfrac{9}{2n} \dfrac{h_o}{S^2} \dfrac{\tilde{K}(h_o)}{f(h_o)}$

   n = coefficient depending on soil's vertical heterogeneity (rate
       slightly less than 1)

In the same way, the discharge tail recession formula is given by:

$$q(t) = \frac{q_o}{(1 + \beta t)^2} \tag{4}$$

where $q(t) = \dfrac{2\gamma}{\alpha} \tilde{K}(h) \dfrac{h^2(t)}{S}$ \hfill (5)

   q(t) = discharge per linear meter drain pipe

   q_o = discharge at the beginning of the tail recession

   γ = second water table shape coefficient (usually $\dfrac{2\gamma}{\alpha} \approx 3.5$)

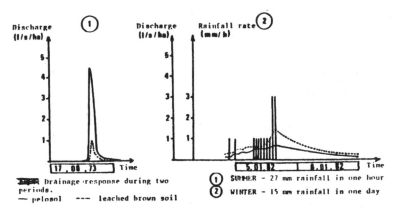

Fig. 4   La Bouzule's drainage hydrographs

The two hydrograph stages are roughly separated by the values q_o and h_o, which enables to draw a simple sketch of a subsurface drained plot (Figure 3), showing the role of the trench.

To end this chapter, two typical hydrographs from La Bouzule are given (Figure 4):

- In winter, the response times to normal rainfalls are short; the peak flow stage is followed by a discharge recession stage, which is shorter in the pelosol. The latter period corresponds to the drawdown of the water table under the tilled layer. During the former period, water mainly flows along the surface and within the tilled layer towards the drainage trench.
- The drainage response to summer storms is instantaneous (high peaks, very short response time). The maximum flow rate is higher in the pelosol than in the brown leached soil. The water either passes down the soil profile through the cracks, numerous and deep in the pelosol, or flows from the surface to the drain pipes through the trench backfill.

4. HYDRAULIC OPERATING CURVES SURVEY THROUGH DOUBLE-MASS METHOD; APPLICA-
   TION TO NETWORKS AGEING'S ANALYSIS

The double-mass method, commonly used in hydrology (Oberlin, 1971), consists of comparing concomittant and sufficiently correlated data sequences:

- on both data sequences X and Y, summations are carried out, so as to obtain two series S_i and T_i ($S_i = \Sigma X_j = S_{i-1} + X_i$; $T_i = \Sigma Y_j = T_{i-1} + Y_i$,

181

X_i and Y_i being the values for each sequence during a given time step);
- the couples (S_i, T_i) are then plotted on an arithmetic graph. If the curve is roughly rectilinear, data are considered as homogeneous. If there is a change in the trend, a rupture in the homogeneity is likely that can be dated and related to a known phenomenon.

Two applications of the method have been performed through data processing (Kinjo et al., 1984). The time step used is a week.

4.1. Definition of drainage seasons

For a given year, the comparison between rainfall and discharge data for each plot enables to accurately separate the drainage seasons. The example of the 1977-78 period in Figure 5 shows that the curves comprise three straight segments, the slopes of which are equal to the season outflow coefficients (defined as the ratio of actual drain flow to rainfall):

- the seasons of drainage beginning (end of autumn) and of drainage ending (beginning of spring) correspond to low coefficients during respectively the wetting up (in autumn) and the drying up (in spring) of the soil;
- during the intense drainage season (middle season) the outflow coefficient is definitely higher because of soil saturation and can reach almost 100%.

The similar shape of the three curves shown in Figure 5 (Arrou, plot 1); La Bouzule, both plots) indicates that, in these different environments,

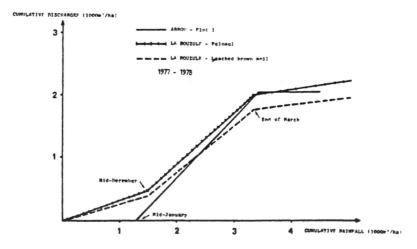

Fig. 5   1977-78 drainage operating curves of three plots

the overall hydrological behaviour is linked with waterlogging. The differences noticed between the outflow coefficients depend on:

a) soil type (cracks in La Bouzule's pelosol may explain the unequal outflow coefficient during seasons 1 and 3);
b) the position in the slope (higher coefficient during season 2 in La Bouzule's pelosol, located downhill from the leached brown soil);
c) construction of drains.

The statistical survey of the dates of the intense drainage season's beginning and ending shows that the former is more variable than the latter (Kinjo et al., 1984).

4.2. Comparison of performance of both plots

The comparison of flow rate data of both plots, either with each other, or with the rainfall data, enables to assess the trends of the hydraulic operating over the eight years, presently available. For example in Figure 6 the discharge versus rainfall drainage operating curves of Arrou's plots 3 and 7, drained with the same installation technique and located in the same way along the slope, puts forward:

- a slight trend alteration from summer 1976 on, as at La Bouzule and in catchment basins (CEMAGREF, 1982). This may be due to the influence of the summer drought in 1976;

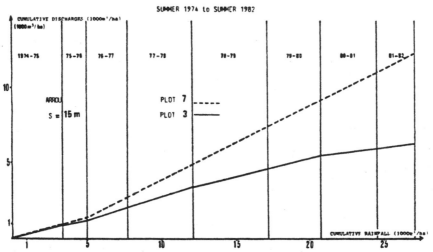

Fig. 6   Arrou's drainage operating curves from 1974 to 1982 (discharge versus rainfall)

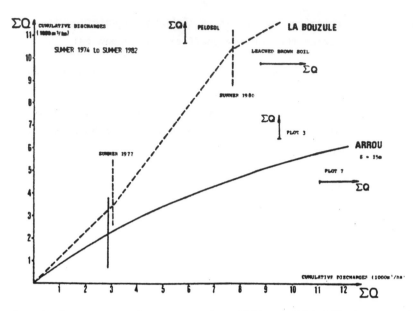

Fig. 7   Drainage operating curves from 1974 to 1982 (discharge versus discharge)

- no trend change for plot 7, which never suffered from bad tillage conditions, but, on the other hand, a decrease in hydraulic efficiency of plot 3.

The analysis of La Bouzule's curves leads to identical remarks.

The application of the double-mass method to flow rate data of both La Bouzule's plots and Arrou's plots 3 and 7 respectively (Figure 7) emphasizes this significant, maybe irreversible, decrease in the hydraulic efficiency of either La Bouzule's pelosol or Arrou's plot 3, which can be related to different causes:

- from summer 1980 on at La Bouzule maize harvests under wet conditions entailed soil compaction and damage to physical properties of the trench backfill (Florentin, 1982; Waleed, 1983). The pelosol network seems to be underdesigned;
- from almost the beginning on, Arrou's plot 3 did not work well, particularly during the year 1979-80, during which Ailliot (1980) noticed a plough pan. It appeared that many pipes were at a too low depth (50 cm) to enable a satisfactory trafficability which leads to soil compaction and likely trench degradation.

184

Even less spectacular drainage operating failures may be spotted through this method, under two conditions:

- fairly long good quality data series must be available, so as to use the method;
- careful investigations of agricultural techniques and of soil and trench profiles, etc. are necessary in order to relate any change in the trend of the curve to the effect of poor trafficability.

Both examples presented here (La Bouzule and Arrou) display networks ageing, which seem to be irreversible. Arrou's plot 3 was repaired in 1984 (installation of new drain pipes at the proper depth); at La Bouzule, tillage has been modified. We shall see within a few years whether the networks have actually been 'renovated'.

5. VALUE AND DURATION OF PEAK FLOWS, APPLICATION TO DRAINAGE DESIGN RATE

The instantaneous peak flow rates depend on numerous parameters (rainfall rate, soil water content, initial physical soil state, etc.). Figure 4 gives an example of different drainage responses of two networks to the same rainfall.

In order to carry out a statistical analysis, the method described by Lesaffre and Laurent (1983) that consists of sampling independent data, which are classified and to which a mathematical curve is fitted, is used. Predictions may be then determined for any frequency (or return period), with a certain confidence interval (here 70%). The method was applied to the two following variables:

- annual or seasonal maximum instantaneous flow rates, which enables to compare land drainage floods to river floods;
- exceedance of given flow rates during a whole year or season, so as to find a design criterion related to the network's lifetime, in a way similar to Rands (1973).

In French drainage practice, calculation of main pipe diameter and of lateral maximal allowable length is based on the drainage design rate, under which networks are not surcharged and which is usually derived from the three days rainfall once a year (except for summer)[*].

---

[*]In France, the design rainfall rate usually ranges from 10 $mm \cdot d^{-1}$ to 20 $mm \cdot d^{-1}$. It can reach 50 $mm \cdot d^{-1}$ in mountaineous areas.

185

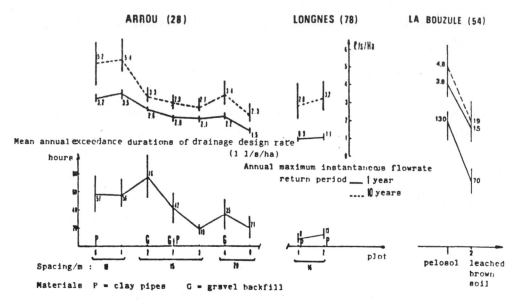

Fig. 8   Peak flow rates (for two return periods) and exceedance durations of design drainage rate. Predictions and confidence intervals
(70%)

Justification of such a design rainfall came from the allowable waterlogging duration, so as to limit the crop production loss (Guyon, 1974).
Accordingly, the drain spacing used to be computed through steady state
formulas. Now, the design rate is not connected to drain spacing, since
the latter is derived from unsteady state formulas (see Chapter 3). On the
contrary, from experience and field observations (Faussey and Hundal, 1980)
it is clear that trench surcharge duration has an effect on drainage efficiency, but the relationship between surcharge duration and efficiency cannot be measured nowadays.

For every plot of the three experimental sites, Figure 8 displays maximum peak flow rates with return periods 1 and 10 years and mean annual exceedance durations of the drainage design rate ($1 \cdot 1 \cdot s^{-1} \cdot ha^{-1}$). Laurent and
Lesaffre (1983) analyzed the effect on these values of environmental
characteristics (plot location, rainfall rate, soil type, drainage season)
and construction systems (drain spacing, trench backfill, drain pipe location, shallow subsoiling). They arrived at conclusions different from those
of Schuch and Jordan (1984), since at Arrou they found that subsoiling had
no effect on peak flows.

From a comparison of the results with farming possibilities on the

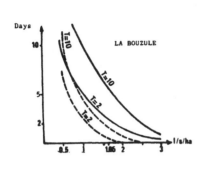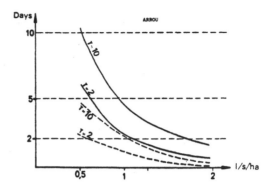

Fig. 9   Discharge exceedance duration (return period: T = 2 years or mean, T = 10 years)
———— pelosol   ----- leached brown soil

drained plots (Ailliot, 1975, 1981; Florentin, 1982), the following conclusions can be drawn:

- following of design and installation rules is a decisive factor for the future performance of the network;
- gravel backfilling does not seem to be useful in Arrou soils, neither from a technical nor from an economical point of view;
- La Bouzule's heavy soil is underdrained, due to a too wide drain spacing (see Section 4.2), and a too low drainage design rate that is exceeded about six days a year, and may explain the trench damage. The usual design rate ($1.65\ 1 \cdot s^{-1} \cdot ha^{-1}$) would be exceeded for sixty hours a year, and a network designed with this rate would probably work better (Figure 9).

6. CONCLUSION AND FURTHER RESEARCH NEEDED

Relevant information can be drawn from the statistical analyses of field experiment hydrographs, when data have been continuously recorded for several years:

- drainage performance is similar in any soil temporarily waterlogged, with differences according to the soil type and construction methods;
- networks ageing (evidence of which can be given through the double-mass method) may be due to either underdesign, installation faults or bad tillage conditions;
- drainage design rates can be derived from allowable trench surcharge duration; the present French approach seems to give a satisfactory compromise.

However, further research in two main directions is necessary:

- hydrograph collecting must be carried on from networks installed in different soil types, under different climates;
- field and laboratory investigations must be conducted on the causes of trench fackfill deterioration, these causes being due to environmental factors and to machine type used.

ACKNOWLEDGMENTS

Thanks are due to Mrs. A. Dumitriu, Messrs J.M. Bouye and A. Mamecier of the CEMAGREF Drainage Division, and to Messrs L. Florentin and A. Kinjo of ENSAIA Nancy (High School of Agronomics).

REFERENCES

Ailliot, B. 1975-1981. Suivi agronomique du réseau d'Arrou, Chambre d'Agriculture d'Eure et Loir. Rapports annuels.

Alessandrello, E., Concaret, J., Guyot, J. and Perrey, C. 1976. Circulation de l'eau en sols limoneux lessivés hydromorphes drainés. C.R. Académie d'Agriculture de France: 364-373.

Al Soufi, R. and Rycroft, D.W. 1975. A method for calculating the effects of mole drainage. FDEU Technical bulletin 751/1:1/7.

Bouzigues, R., Favrot, J.C. and Hallaire, V. 1981. French heavy soils; characterisation and cartography in relation to drainage. Land Drainage Seminar, Cambridge. Balkema, Rotterdam: 33-53.

CEMAGREF. 1982. Analyse des influences du drainage agricole sur l'hydrologie de quelques petits bassins-versants de l'Ile de France: examen de quelques séries chronologiques et enseignements à tirer. Rapport provisoire.

Devillers, J.L. and Guyon, G. 1978. Choice of a field drainage treatment. International Drainage Workshop. ILRI 25: 165-179.

Faussey, N.R. and Hundal, S.A. 1980. Role of trench backfill in subsurface drainage. A review. Trans. ASDE 23-5: 1197-1200.

Florentin, L. 1982. Contribution à la connaissance des sols hydromorphes et apparentés de Lorraine, et de leurs réponses au drainage. Thèse Doctorat, INPL Nancy 90-123.

Guyon, G. 1961. Quelques considérations sur la théorie du drainage et premiers résultats expérimentaux. BTGR no. 52. Ministère de l'Agriculture. 44 pp.

Guyon, G. 1966. Considérations sur l'hydraulique du drainage des nappes. BTGR no. 79, Ministère de l'Agriculture.

Guyon, G. 1974. Le drainage agricole. Essai de synthèse. BTGR no. 117, CEMAGREF: 1-66.

Guyon, G. and Wolsack, J. 1978. The hydraulic conductivity in heterogeneous and anisotropic media and its estimation in situ. Proc. International Drainage Workshop. ILRI 25: 124-135.

Guyon, G. 1980. Transient state equations of water table recession in heterogeneous and anisotropic soils. Trans. ASAE 23-3: 653-656.

Guyon, G. 1983. Le périmètre expérimental de drainage d'Arrou. Aspects hydrauliques. Etudes du CEMAGREF. Hors Série no. 5. 92 pp.

Herve, J.J. 1980. Limites et validité des modèles hydrodynamiques. Document CEMAGREF. 20 pp.

Kinje, A., Lesaffre, B. and Morel, R. 1984. Restitution et courbes de fonctionnement hydraulique en drainage agricole. Académie d'Agriculture de France 70-2: 278-288.

Laurent, F. and Lesaffre B. 1983. Etude statistique des débits élevés en drainage agricole. Etudes du CEMAGREF Hors série no. 6. 122 pp.

Lesaffre, B. and Laurent, F. 1983. Le fonctionnement hydraulique des réseaux de drainage agricole: débits de pointe et de tarissement non influencé, rôle de la tranchée de drainage. Académie d'Agriculture de France: 1167-1178.

Oberlin, G. 1971. Généralités sur les exigences et contrôles de qualité des données hydrologiques de base. Note interne CEMAGREF.

Rands, J.G. 1973. An analysis of drainflows from FDEU experimental sites. FDEU Technical Bulletin 73/11. 15 pp.

Russel, J.L. 1934. Scientific research in soil drainage. Jour. of Agric. Sci. 24: 544-573.

Schuch, M. and Jordan, F. 1984. Experience on water management over 12 years in fields with amelioration in three grades. Colloque AFES, Dijon. AFES: 217-228.

Trafford, B.D. 1983. The relationship between field drainage and arterial drainage. Theoretical aspects. FDEU Technical Bulletin 73/10. 14 pp.

Van Hoorn, J.W. 1973. Drainage of heavy clay soils. Drainage principles and applications. ILRI no. 16, vol. IV: 313-326.

Van Schilfgaarde, J. 1963. Design of tile drainage for falling water table. Journal of the irrigation and drainage division. Proc. ASCE 89 (IR. 2): 1-12.

Van Schilfgaarde, J. 1965. Transient design of drainage systems journal of the irrigation and drainage division. Proc. ASCE 4458 (IR. 3): 9-22.

Waleed, J. 1983. Comportement hydrodynamique des sols lourds lorrains drainés en fonction des technologies de drainage. Thèse INPL, Nancy.

Design, installation and maintenance of drainage systems

H.RADERMACHER
Universität Bonn, FR Germany

ABSTRACT

The paper starts with a general view of the planning process in connection with drainage systems (design, installation, maintenance). After
mentioning reasons, aims and methods of drainage, a summary of design parameters for open and closed drains is given. In the following, installation and
operating of drainage systems is considered. A special chapter deals with
the maintenance of drainage networks. On the basis of information about the
different methods of maintenance, an analysis concerning the connection between necessary working processes for fulfillment of maintenance aims and
the maintenance method is made. This enables the deduction of an evaluation
of maintenance methods.

1. INTRODUCTION

The planning of drainage systems must be considered as a complex matter,
involving a lot of parameters and interdependencies between them. Therefore
it makes sense to divide the whole problem into separate steps. According to
figure 1 one may distinguish generally between strategic and concrete planning levels. The strategic planning level contains the taking over or the
formulation of higher level aims (that means for example the decision about
the priority of agricultural or ecological function of the region under consideration).

The level of concrete planning may be subdivided into the following
steps:

- detailed analysis of the situation in the area;
- transferring of the general strategic aims to the specific situation (i.e.
  formulation of concrete aims for the area, for example decision on artificial drainage or not and the desired groundwater level considering the
  function of and conditions in the area with regard to agricultural and
  ecological aspects);
- design of the drainage system, i.e. determination of all parameters which
  are necessary to construct the system. These parameters may be considered
  as a function of natural resources and circumstances, given aims and technical possibilities;
- installation of the drainage system;

191

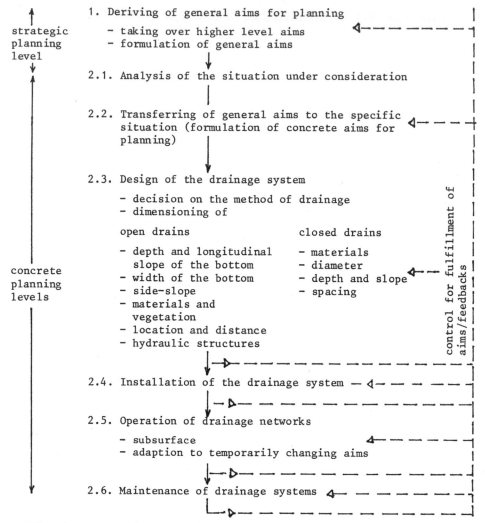

Fig. 1   Generalized planning scheme of drainage systems

- operation of the drainage network, i.e. manipulating water levels;
- maintenance of the drainage system.

The described listing cannot be regarded as an exact chronological scheme, but it differentiates the planning process into steps. The interdependencies between parameters in each step (resulting parameters, factors of influence) to those of others necessitate feedbacks. Besides there must be control for fulfillment of aims after each step. Therefore the scheme should be used in an interactive procedure until the results fulfill the criteria

192

set for the formulated aims. In this paper more attention is paid to making clear the general interdependencies dealing with design, installation and maintenance of drainage systems than to giving details.

2. REASONS FOR WATERLOGGING, AIMS AND METHODS OF DRAINAGE

After having discussed the reasons for waterlogging, this chapter conprehends the aims and methods of drainage.

2.1. Reasons for waterlogging

Considered from an agricultural point of view, drainage becomes necessary, if the following situations occur:

- too high level of groundwater;
- harmful amount of water in the root zone caused by a layer of very low permeability near to the soil surface;
- too big amount of suspended water;
- water ponding on the soil surface during too long times.

These conditions resulting for example in yield reduction of cultivated crops may be caused by the following factors:

- soil characteristics;
- precipitation and snowmelt;
- flood or surface flow from other areas;
- groundwater inflow;
- irrigation;
- water level in the outlet ditches or regional outlet channels.

The mentioned causes for waterlogging are worth to notice, because they should be taken into account in connection with the determination of type, structure and dimension of the drainage system.

2.2. Aims of drainage

Concerning the aims, it is important to distinguish two different situations:

- drainage of larger areas: control of soil water in order to fulfill given aims of the area (i.e. for example to achieve a groundwater level which enables contribution of capillary rising water to the water supply of the

crops without exercising harmful influence on the yield by other effects
such as salinization);
- drainage of special objects (in points, linear) drain off of harmful water
from the object.

In the following, mainly attention is paid to the drainage of areas with
special regard to their agricultural or ecological function.

2.3. Methods of drainage

Ignoring natural drainage, the following methods are available to
achieve drainage (see figure 2):

- open drains
- closed gravity drains
- drainage pumping
- well drainage
- subsoiling methods
- combination of above mentioned methods

In the following the paper only deals with drainage by open and closed
drains. Concerning closed drains there is no further notice taken of mole
drainage. Because of the close connection between aims, function and method
of drainage, the main functions of the drainage methods under consideration
have been summarized as follows:

- open drains:
  . providing of surface drainage (floods, strong precipitation, irrigation);
  . subsurface drainage;
  . serving as outlets for closed drains.

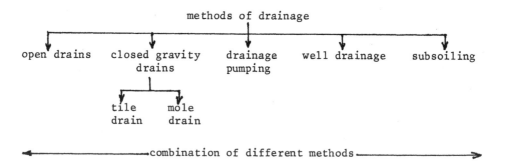

Fig. 2   Methods of drainage

194

- closed drains:
  . control of groundwater level;
  . carrying away of water percolated through the soil (for example caused by irrigation losses or leaching to reduce salt concentration in the root zone).

3. DESIGN OF DRAINAGE SYSTEMS

This chapter gives a concise overlook about the design of drainage systems by mentioning the parameters which are necessary to construct and operate the system.

3.1. Choice of the drainage method

The choice of the kind of drainage method depends upon an evaluation of the following factors with regard to the special situation to be treated:

- reasons for waterlogging;
- aims and extent of the intended measure;
- characteristics of open and closed drains (figure 3 gives advantages and disadvantages of open drains as compared with closed drains);
- economics.

Advantages	Disadvantages
- quick drain off - big hydraulic capacity - applicable under unfavourable soil conditions - small longitudinal slope sufficient - easy to construct - low cost for construction - easy to modify (for example application of subsurface irrigation) - quick detecting and repairing of damages	- large loss of land - handicap for use of agricultural machinery - high maintenance costs - increased danger of weeds and insects coming from the open drains - reduced functioning during frost

Fig. 3   Advantages and disadvantages of open drains as compared to closed drains

3.2. Dimensioning of drainage systems

In this connection, the expression dimensioning should include the determination of all parameters required for construction of open and closed drainage systems (see figure 1). The dimensioning has to be carried out in such a way that the planned network fulfills the formulated aims under the given conditions. In the following the above introduced division of drainage systems in open and closed networks will be used.

Dimensioning of open drains

The dimensioning of open drains includes the determination of the following parameters:

- depth and longitudinal bottom slope
- bottom width
- side slopes
- materials or vegetation in the cross-section
- location and distance of open drains
- hydraulic structures (also in connection with the use of system for subsurface irrigation)

Correct dimensioning means determination of values concerning the above mentioned parameters in such a way that the functioning of the system (discharge and water level) is in accordance with the formulated aims also with regard to interdependent planning steps, for example: maintenance planning.

Dimensioning of closed drains

The dimensioning of closed drainage networks contains the fixing of following parameters:

- materials
- diameter
- depth and slope
- spacing

Concerning the connection between the determination of dimensioning parameters and the functioning of the system, the statements mentioned above are also valid.

Because of its importance, in the following some information dealing with materials is given:

- in former times (until 1965) clay tile was by far the most widely used material for closed drains;
- concrete tiles of special quality (manufactured for example from sulphate

resistent cement) are designed for use in case corrosive acids or sul-
phates are present in soil or water;

- plastic tubing now seems to replace the other materials rapidly because
of its low cost and easy installation.

4. INSTALLATION OF DRAINAGE SYSTEMS

In this connection the installation or construction of drainage systems
may be defined as transferring the determined design parameters to reality.
In accordance with the above introduced division of drainage networks the
following description is given separately for open and closed systems.

4.1. Construction of open drains

The construction of open drains may be realized by various types of
equipment including dragline, scraper, backhoe, grader, bulldozer, wheel-
type excavator, shovel and plow. According to the degree of accuracy required, an adequate surveying method must be applied in order to meet the
design aims concerning place and level.

4.2. Installation of closed drains

With regard to the way of installation it is possible to distinguish
between two different procedures:

- digging a trench and putting the closed drain inside;
- installation of closed drains without using trenches. This method is commonly applied since continuous plastic tubing is available.

The slope of the closed drain must be realized in the first case by
using leveling instruments; in case of trenchless installation, laser
equipment is available. Closed drains commonly are used in connection with
envelopes to achieve the following advantages:

- preventing of soil particles entering into the tubing;
- establishing of an zone of improved permeability around the drain;
- stabilization of the tubing.

Envelopes may consist of:

- organic materials (hay, straw, sawdust, wood chips, peat moss, maize
cobs);

197

- anorganic materials (sand, gravel, cinders, slag, crushed stone);
- synthetic fabrics (nylon, polyester, polyethylene).

5. OPERATING OF DRAINAGE SYSTEMS

The fulfillment of the function of the drainage network must be realized in the design and construction and should be preserved by maintenance. In this connection operating includes activities to influence discharge in the drainage system and therewith the groundwater level of the area by manipulating the water level at special points in the drainage network. This may be done with regard to the following purposes:

- rising of the groundwater level in order to contribute to the water supply of the plants (subsurface irrigation);
- adaptation of the groundwater level to temporarily changing aims.

6. MAINTENANCE OF DRAINAGE SYSTEMS

Before dealing with details, it appears to be helpful to clarify some terms:

- Preventive maintenance; operations performed in preserving drainage canals and hydraulic structures in good or near-original condition to ensure their planned function. In order to preserve the fulfillment of aims by preventive maintenance only, it is necessary that the planned condition of the drainage system is existing. If not, the necessity of renewals is arising.
- Renewals or corrective maintenance; extensive repairs of drainage canals and hydraulic structures, including restoration and replacement of damaged portions of the considered objects.

6.1. Maintenance of open drains

The following facts make maintenance necessary:

- change of the cross-section and longitudinal slope caused by sedimentation and erosion;
- growing of weeds in the drainage system.

The mentioned facts mainly result in a decreasing hydraulic capacity of the network and rising water levels in the system and therewith rising groundwater level so that the planned function is not guaranteed. In order to preserve the planned condition of open drains, the following methods of maintenance may be applied:

- physical weed control (weed burning);
- biological weed control (like shading of drains by trees, introduction of weed eating fish, or setting of special plants to prevent weed growth);
- mechanical maintenance (cutting and removing of weeds, clearing away of sediment);
- chemical weed control, i.e. applying chemicals which by themselves or with their adjuvants kill weeds.

Figure 4 gives a general view of the working processes which are necessary for complete fulfillment of maintenance aims (mowing, killing or preventing of weed, removing of weed, taking away of sediments). As this figure shows the following statements may be derived for the connection between these working processes and the above mentioned methods of maintenance:

1) mechanical maintenance is the only method, which realizes all required working processes;
2) chemical weed control entails negative side-effects:

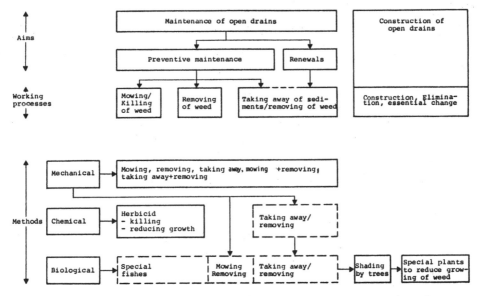

Fig. 4   General view of maintenance methods and necessary working processes (with regard to the aims)

- toxic effects;
- weed may get resistant to applied chemicals (in this case it becomes necessary to add or change to other chemicals to kill the secondary flora; this causes higher costs);
- rising risk of damage be erosion because the stabilizing effect of the plants is vanishing;
- possibility of decreasing content of oxygen in the water as a result of the desintegration of the killed organic material;

3) the physical weed control may entail the two last mentioned negative effects of 2;

4) experiences show that biological weed control alone seems not to be sufficient to maintain drainage systems (biological measures may be considered as part of the design and construction).

The result of the analysis of maintenance working processes and methods may be summarized by the following statements:

- from an ecological point of view the mechanical maintenance is the most favourable method;
- long-term consideration also proves the advantage of mechanical maintenance with regard to economics

6.2. Maintenance of closed drains

The function of closed drains may be disturbed by sediments, roots and .chemicals. In order to avoid damage of closed drains, the following measures can be used:

- physical method (high pressure water-jet cleaning);
- chemical treatment (applying of sulfur-dioxide to dissolve oxide deposits).

6.3. Final conclusions

The above mentioned facts enable the deduction of some important statements:

- as stated in the introduction, there exist interdependencies between all steps of planning. This makes it necessary to realize each planning step in connection with other steps. Therefore, the achieving of results for each step includes a feedback and a control for the fulfillment of aims of other steps (iterative procedure of planning). This can be illustrated

200

by two examples in connection with maintenance:

. because both maintenance and design influence discharge and water level
  in the drainage system, it is necessary to coordinate the parameters in
  the respective planning steps in such a way, that both aims are fulfilled;

. since the method of maintenance may affect ecological conditions, the
  evaluation of maintenance methods must consider the function of the
  drainage system and the area (for example, chemical weed control becomes
  very problematic, if water is used for drinking or irrigation);

- the responsibility for planning, financing and carrying out the required
  maintenance should be prescribed by rules and taken over by special orga-
  nizations. In the Federal Republic of Germany the setting of legal rules
  concerning maintenance is in hands of the federal state. It should be
  pointed out that these rules are differentiated strongly with regard to
  the function, but they often neglect ecological aspects;

- until now the ecological aspect is hardly considered in design of drainage
  systems. In future planning and maintenance more attention has therefore
  to be paid to ecological aspects. The coordination of agricultural and
  ecological aims requires a splitting up into regions in which the priority
  of the mentioned aspects is fixed.

7. REFERENCES

Achtnich, W. 1980. Bewässerungslandbau. Verlag Eugen Ulmer, Stuttgart. 199
    pp.
Baitsch, B. und Radermacher, H. 1968. Gewässerunterhaltung. Schriftenreihe
    des Kuratoriums für Kulturbauwesen.
Jensen, M.E. 1980. Design and operation of farm irrigation systems. The
    American Society of Agricultural Engineers, St. Joseph, Michigan.
Radermacher, H. 1978. Arbeitsblätter zur Wasserwirtschaftlichen Planung I.
    Lehrstuhl für landwirtschaftlichen Wasserbau und Kulturtechnik der Uni-
    versität Bonn.

The effect of maintenance on the performance of tube drains

G.A.VEN
IJsselmeerpolders Development Authority, Lelystad, Netherlands

ABSTRACT

In the newly reclaimed areas in the IJsselmeer district and the Lauwers-zee pipe drainage systems are necessary to control the groundwater table. The main problem with drainage performance is clogging of the drains by iron ochre or sulphur deposits. Drainage research on envelope materials is done with sandtanks and on experimental fields. Some results of measurements on experimental fields related to drainage maintenance are discussed.

It could be concluded that clogging by iron ochre results in an increased entrance resistance and in a reduction of discharge capacity of the drains. Flushing of the drains was not effective in reducing the entrance resistance. The discharge capacity can be recovered by flushing. Therefore the lifetime of the drainage system is mainly determined by the rate of increase of the entrance resistance.

When there is a risk for iron ochre formation, fine structured non-voluminous envelopes are not suited.

1. INTRODUCTION

The IJsselmeerpolders Development Authority (RIJP) is responsible for the development of newly reclaimed areas in the IJsselmeer district and the Lauwerszee (Figure 1). The project to reclaim parts of the IJsselmeer started in 1927. Since then 4 polders have been reclaimed and developed with a total area of 165 000 ha. Agriculture is the most important type of land use. Recently the development of nature reserves, forests, recreational and urban areas became more important.

The top layers of the soils in the IJsselmeerpolders usually consist of marine clay and loamy deposits containing 12 to 40% clay and ranging in thickness from 0 to about 8 m. Below these Holocene sediments sandy Pleistocene deposits are found. The texture is generally very fine to medium fine. For most types of land use the installation of tube drains is necessary. At places where the drains have to be installed in sandy layers, the application of envelope materials is necessary. In clay soils the drains are installed without envelopes.

In the Lauwerszee part of the area will be used for military exercises. Installation of drains is necessary to control the groundwater table. The soil type in this area is marine fine to medium fine sand. Envelope materials are necessary to prevent silting up of the tubes.

Fig. 1   Location of the IJsselmeerpolders and the Lauwerszee

2. DRAINAGE PROBLEMS

The main problem with drainage performance in the IJsselmeerpolders and the Lauwerszee is clogging of the drains by:

- invasion of soil particles
- root growth
- iron ochre and sulphur deposits

Clogging of drains by soil particle invasion is a minor problem. The last 25 years continuous practical research has been done to select proper envelope materials (Scholten and Ven, 1984). Based on the results of this

204

research envelope materials for the local situation can be selected with a low risk for silting up and acceptable low entrance resistances.

Special in forests clogging of the drains by roots is a common problem. The most severe problems have been noticed in loamy to sandy areas with a compacted subsoil. The roots preferably penetrate in the less compacted drain trench, which is often richer in nutrients because of the mixing. In some forests in Eastern Flevoland drains were blocked completely 5 to 10 years after installation. Up to now there is no suitable method to remove these roots and to reactivate the drains (Slager, 1981). The best method to increase the lifetime of the drains is to flush them frequently.

Clogging of the drains by iron ochre formation is the main problem in the IJsselmeerpolders. In clay soils oxidation of iron takes place in the profile itself, which can be seen from the change in colour from blue-grey into brown. In and around the drain ochre formation does not occur. Severe problems can arise on sandy soils and on clay soils in urban areas where the surface level has been raised artificially with about 1 m of sand by means of hydraulic landfill. The source of the iron may be autochtone or allochtone (seepage of groundwater rich in iron). The performance of the drainage system on the long run is mainly determined by the impact of ochre formation and on maintenance practice.

In the Lauwerszee area the quality of the groundwater resembles that of seawater. Especially near the seadike there is continuous seepage from the sea. Drains in pilot areas became completely blocked within 5 years by iron and sulphur deposits. Attempts to recover the drains by flushing were not successful.

3. DRAINAGE RESEARCH

Drainage research at the RIJP is done by means of sand tank experiments and on experimental fields. In sand tank experiments local types of sand are used and natural flow conditions are simulated as good as possible. Each experiment with a combination of soil type and drain envelope takes 3 weeks. From the measurements the entrace resistance of the drain and the amount of sand washed into the drain is determined. In this way the performance of different materials can be compared and envelope materials with a low entrance resistance and a low risk of silting up can be selected for use in practice.

From these laboratory experiments no information can be obtained about the durability of envelopes and about the performance on the long run, especially when chemical clogging danger exists. Therefore also field experiments have been set up to test envelope materials under practical conditions. In these field experiments the following variables are monitored:

- drain discharge
- groundwater table midway between drains
- hydraulic head in the drain trench
- hydraulic head in the drain pipe

From these data again the entrace resistance of the drain can be derived. The hydraulic head in the drain pipe itself gives an indication about the discharge capacity. Because of ochre formation drains have to be maintained, hence the effect of flushing on drainage performance is also studied.

In the Lauwerszee research has been performed to select proper envelope materials for the local situation. Because regular maintenance will be necessary after installation, field experiments have been done with different types of flushing machines.

In the next part of this paper some results of field experiments on drainage maintenance will be discussed.

4. RESULTS OF FIELD EXPERIMENTS

The RIJP has several experimental fields where envelope materials are tested. A review of the results of these experiments has been reported by Scholten (1985). In this paper only some results related to chemical clogging and effects of maintenance will be discussed. It should be noted that part of the results were obtained from experiments in areas not in use for agriculture, but the results are generally applicable to drained agricultural land too.

4.1. Experimental field B49

This field is situated near Lelystad. Because the area was meant for housing the original profile of heavy loam has been covered with 1 m of coarse sand. Drains have been installed in 1975 with the following specifications:

- material: corrugated PVC 65/58 mm
- spacing: 24 m
- drain depth: 1.90 m
- drain length: 250 m

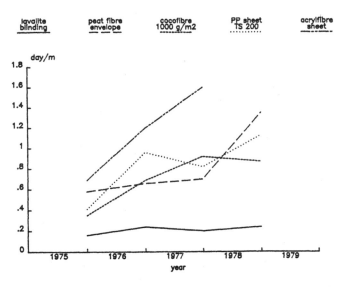

Fig. 2   Entrance resistances of different envelope materials on plot
B49

- backfill trench: coarse sand mixed with heavy loam
- envelope materials: . glass slag
                      . lavalite grade 3-6 mm (a vulcanic gravel)
                      . coconut fibre 1000 g/m^2
                      . PP sheet TS200, 3 mm thick
                      . acrylfibre sheet, 3 mm thick

Measurements were carried out during a number of consecutive winter periods. Groundwater tables and hydraulic heads inside the drain and in the
trench were done at 100 m from the outlets. The results about the entrance
resistance as a function of time are indicated in Figure 2. Because of ochre
formation the entrance resistance increases with time, except for granular
envelope materials. The drains were flushed with a high pressure pump in
November 1978. The entrance resistances after flushing have not been lowered by this operation.

The mean discharge during wet periods decreased gradually with time.
After flushing the same mean discharge as in the first winter period could
be realized. The effect of ochre formation on the discharge capacity of the
drains is illustrated by Figure 3. In this figure the mean hydraulic heads
relative to the bottom of the drain are plotted versus time. Within 5 years
the hydraulic head in the drain increased up to 0.9-1.0 m above drain level,

207

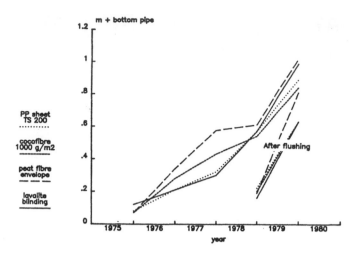

Fig. 3   Mean hydraulic heads in the drains on plot B49

resulting in high groundwater tables during wet periods. By flushing the discharge capacity could be increased to an acceptable level.

4.2. Experimental field P106

   This field is situated in Eastern Flevoland and is in use as arable land. The top layer of 0.30–0.50 m is heavy loam. Underneath a layer of 0.10–0.40 m of peat is found. The subsoil is medium fine sand. Because of the high transmissivity of the underlying aquifer and the location adjacent to the polder dike there is a strong seepage. The seepage water is rich in iron, so the drains have to be cleaned regularly.

   The first drains on this plot have been installed in 1961. Later on the experiments have been extended with other drainage materials. Characteristics of the drainage system are:

– material: mainly corrugated PVC 50/44 mm
– spacing: 8 m
– drain depth: 0.9 m
– drain length: 100 m
– envelope materials: several

   Measurements of hydraulic heads and groundwater tables are done at 80 m from the outlet.

   Concerning the entrance resistance similar conclusions as from plot B49 can be drawn. Envelope materials giving strongly increasing entrance resis-

208

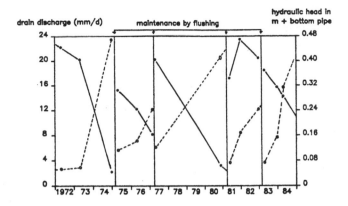

Fig. 4   The effect of flushing on the discharge rates and the hydraulic head on plot P106 (object coconut fibre)

tances with time were synthetic non-voluminous fabrics as glass fibre, PP sheet and also PS granules in a perforated plastic foil. For materials like coconut fibre, fibrous peat and PP fibre gave an acceptable low entrance resistance 5 to 10 years after installation. The entrance resistances could not be reduced noticeably by flushing.

With respect to recovering the discharge capacity, flushing was more successful. As an example the mean discharge and hydraulic head inside the drain pipe of drains wrapped with coconut fibre are plotted in Figure 4. Because of the strong seepage the discharge with proper functioning drains should be more or less constant on a level of 20 to 24 mm/day. Without maintenance the discharge decreased strongly within a few years, while the hydraulic head increased. Also the groundwater table midway between drains is affected negatively by blockage of the pipes.

From inspection of the drains it proved that most of the ochre formation in the pipes occurs in the last 10 m from the outlets. Regular maintenance is inevitable to keep the drains functioning.

4.3. Experience in the Lauwerszee

In the northern part of the Lauwerszee an area of about 1600 ha had to be drained, to make it suitable for military exercises. The type of soil is sand with an average grain size D50 varying from 90 to 130 µm and a clay content between 1 and 5%. The groundwater is still saline because of the reclamation from the sea in 1969. Near the sea dike there is seepage from the sea.

Because problems were expected with clogging of the pipe drains by silting up and by iron and sulphur deposits a pilot area has been drained in 1970. After 5 years the envelopes and the perforations of the drains were blocked completely. Attempts to recover the drains by flushing with high pressure (80 bars at the pump) failed.

In 1976 again new drains have been installed. To study the possibility to increase the lifetime of the drains, the laterals were flushed each year with high pressure equipment. In the third year problems arose because part of the drains could only be cleaned up to half of their length. It was also very difficult to draw the hose out of the drain pipe. The reason of the problem was silting up, caused by the repeated flushing operations. This was especially the case for drains wrapped with coconut fibre and for drains blinded with fine gravel. Drains with a peat fibre envelope were almost free of sand. A solution for preventing silting up was to apply gravel above and below the drain. Peat fibre was less attractive because of the higher entrance resistance caused by this type of material.

In following experiments it was found that drains enveloped with coconut fibre (1000 g/m^2) or PP fibre (8 mm thick) could be flushed safely with medium (40 bar) and low pressure (20 bar). The cleaning effect with medium pressure was somewhat better than that with low pressure (Brinkhorst et al., 1983).

5. CONCLUSIONS

For tube drains in sandy to loamy soils with a risk of clogging by iron ochre formation, the following conclusions can be drawn with respect to the choice of envelope materials and maintenance:

- Clogging by ochre formation results in an increase in entrance resistance and a reduction of the drainage capacity of the drains.
- Flushing of the drains is not effective in reducing the entrance resistance. The discharge capacity can be recovered if the blockage is mainly caused by ochre.
- The lifetime of the drainage systems is mainly determined by the rate of increase of the entrance resistance in time.
- Fine structured non-voluminous envelopes are not suited for application under those conditions. The best choice is to apply voluminous coarse structured materials.
- Flushing with high pressure can be dangerous because it creates an unstable situation around the pipe. Under these conditions extreme silting up may occur.

REFERENCES

Brinkhorst, W., v.d. Linde, K. and Scholten, J. 1983. Ervaringen met het doorspuiten van drains in het Lauwerszeegebied. RIJP rapport 1983-6 abw.

Scholten, J. and Ven, G.A. 1984. Drainage materials survey in the IJssel-meerpolders. RIJP rapport 1984-28 abw.

Scholten, J. 1985. Onderzoek aan omhullings- en afdekkingsmaterialen over de jaren 1978 tot 1984. RIJP rapport (in preparation).

Slager, H. 1981. Functioneren en onderhoud van ontwateringsstelsels in bossen in Flevoland. Flevobericht no. 180, RIJP, Lelystad.

Discussion Session 3

Question from Mr. A.L.M. van Wijk to Mr. W. Dierickx

My question has nothing to do with envelope materials. Why did you apply
such shallow drain depths of 0.60 and 0.70 m in your experimental
fields?

Answer of Mr. Dierickx

These shallow drain depths were necessary because of the water level of
the evacuation ditches. Only limited financial means were available to
install drain pipes on a field suitable for that purpose. In fact the
whole dewatering system had to be improved but the financial means were
insufficient for that. Therefore we decided to install the drains at a
depth shallower than the commonly used depth of 0.80 to 1.00 m.

Question from Mr. A. Puraye to Mr. L.C.P.M. Stuyt

Do you have any knowledge about the clogging of PVC pipes by calcareous
deposits. The problem occurs at different locations in Luxemburg. What
measures can be taken to prevent the formation of this type of crusts
and is removing of them possible?

Answer of Mr. Stuyt

We do not have experience with the clogging of calcareous soils.

Answer of Mr. Dierickx

Internal flushing of the pipe does not help in these cases. It is a very
serious problem and I do not know how to solve it. We found a road
drainage, where even a gravel envelope was completely blocked by
calcareous deposits.

Answer of Mr. Lesaffre

In France, we are of the same opinion. Fortunately, this type of clogging is very rare.

Question from Mr. A. Armstrong to Mr. W. Dierickx and Mr. L.C.P.M. Stuyt

In the UK we have also been looking at drainage filters, and my colleague
Mr. Dennis has concluded an investigation into the establishment of a
technical standard for such materials. This work has centred on the
determination of the opening size distribution in the filter material,
and the relation between desirable opening sizes and the possible sizes
in the surrounding soils. I was thus interested in your observation that
the material in one pipe you showed us was largely clay, implying a
differential movement of soil components. An experience, derived from

213

a nationwide survey of siltation in drain pipes, is that the material
found in pipes is very identical to that in the surrounding soil, and
there was no evidence of differential soil particle movement. Would you
like to comment on this aspect?

Answer of Mr. Diericks

From previous experiments with cases where no envelope materials were
used or in cases where something was going wrong with the envelope
we also found little or no difference in particle size distribution between the sediments found in the pipe and that of the soil abutting the
drain. The research described in this contribution just stresses the
possibility to predict the need of an envelope material by laboratory
research. Once the need has been established, the kind of envelope to
use has to be determined. Therefore just an evaluation about the rate
of sedimentation of the drain pipes was considered to find the effectiveness of the various treatments or envelopes was performed without
further analysis.

Answer of Mr. Stuyt

We have also done analyses of the textural composition of the particles
in the drain pipe. We generally observe that the coarser particles remain in the pipe and that smaller particles remain suspended and are
washed out subsequently. One should be aware of the fact that the
particles in the drain can be washed out at various places in the soil.
In lighter soils the particles may be washed out from the trench backfill; in heavier soils however, the particles may come from remote
places like the soil surface, and are transported via the soil cracks,
like was found in France. In our laboratory set-up, particles that are
washed out from the soil are largely retained since we don't have the
high drain flow rates that occur under field conditions. Therefore, we
can compare particle size distributions and the filtering action of
envelopes more precisely in the laboratory.

Question from Mr. L. Galvin to Mr. B. Lesaffre

Your figure 5 shows the three flow segments for 1977/78. In figure 6
there are no indications of segmented flow. Is this because only annual
totals are plotted in figure 6?

Answer of Mr. Lesaffre

You are right. The curves have been smoothed, so as to clearly display
the interannual hydrological trend. Otherwise, they would have consisted of steps.

Question from Mr. J. Wesseling to Mr. B. Lesaffre

The Double Mass method does not give information on the intensity of the drainage system. In this respect one has more to rely on the exceedance curves. What is your opinion about this?

Answer of Mr. Lesaffre

The exceedance curves give only information on the occurrence of high flow rates, but these flow rates may occur any time, for example immediately after the network has been installed or ten years thereafter. Moreover, from a theoretical point of view, one must check that every flow rate has the same meaning over years, i.e. that there is no change in the hydrologic trend. The Double Mass method does not provide detailed information on the intensity of the drainage system, but provides relevant information on its possible change. Finally, I should like to emphasize that it should be interesting to apply this rather simple method to data available in other countries.

Question from Mr. Ph. Lagacherie to Mr. G.A. Ven

In Languedoc we had iron clogging. We observed it was linked with depth of drain (drain in a peaty layer or not) and was more or less influenced by water table. In the same area we observed that a large drain slope has a good effect in the prevention of iron clogging. Have you observed the same phenomena?

Answer of Mr. Ven

The deeper the drain, the more unaerobic conditions occur. This is favourable, because iron clogging can be reduced when there is less oxygen available. Still the problem cannot be solved completely by installing drains deeper.

Concerning to the second question: we have no experience about the effect of the drain slope in iron clogging. I can only say that the problems increase when iron clogging is combined with mineral clogging. Special when soil/iron mixtures become dried up, the sediments are difficult to remove by flushing. Because an increasing slope reduces mineral sedimentation it may be effective. In the Netherlands large slopes are not possible because of the flat topography.

Question from Mr. L. Galvin to Mr. G.A. Ven

After each flushing you achieved a substantial reduction in hydraulic head. The first two flushings generated an increase in drain discharge followed by a gradual decrease as the hydraulic head increased. How-

ever, the maximum flow rate did not occur immediately after the third
flushing, but later (about January 1982). Can you explain this?

Answer of Mr. Ven

The maximum flow rate is also dependent on factors other than the performance of the drains, e.g. the water level in the lake outside the
polder which effects the seepage rate. Because this level has a seasonal fluctuation also the seepage rate varies. There is also an effect
of the weather conditions (wet, dry periods) on the discharge. As each
point in the graph is an average of a number of observations there is
some bias in the average discharge. The anomality after the third
flushing is probably the result of a combination of the above mentioned effects and not caused by the condition of the drain.

Question from Mr. W. Dierickx to Mr. G.A. Ven

You said that you do not have problems with envelope materials. For
sandy soils I can agree with your opinion because there you can use
every envelope material if not too coarse. But there are a lot of other
soils which need envelopes around the drains or for which it is difficult to decide upon the need of an envelope material.
Additionally, why do you apply flushing if you don't have sedimentation
problems?

Answer of Mr. Ven

In the IJsselmeerpolders the selection of envelopes is not the main
problem. Fine structured envelopes are not useful because the need for
enveloping goes always hand in hand with the risk of iron-clogging.
The selection is more a problem of costs and quality of the envelope.
Gravel of good quality gives very good results, but is too expensive
for agriculture. A good second choice is a mixture of PP fibre of good
quality. In very fine sands with high risk of mineral clogging peat
fibre is successful. Always the quality of the envelope is important.
In answer to your second question: flushing is mainly done to clean
the drain internally from ochre and other chemical sediments.

Question from Mr. S.E. Olesen to Mr. G.A. Ven

You conclude that the buildup of radial resistance cannot be stopped
by flushing. In Denmark we have the experience that, in some cases, we
can reduce radial resistance by flushing, however.

Answer of Mr. Ven

On both our pilot areas we come to the same conclusion: we cannot reduce the entrance resistance by flushing. The only action of flushing
is that it cleans some perforations of the pipe wall internally.

Answer of Mr. Stuyt

The Dutch Polder Development Authority and the Institute for Land and
Water Management Research (ICW) are planning to study these phenomena
in more detail. In a field situation, pressure transducers will be installed in the vicinity of the pipe, with which the water pressure
buildup can be monitored when flushing the pipe. In doing so, we will
be able to study how far the influence of flushing penetrates into the
soil.

Answer of Mr. Lesaffre

From our experience in France, it seems that physical iron clogging increases the entrance resistance, whereas biological iron clogging reduces the transport capacity of the drain. Flushing has a better effect
in the latter case.

Question from Mr. S.E. Olesen to Mr. G.A. Ven

You concluded that flushing of iron-clogged drains did not reduce the
entrance resistance. This is not in agreement with Danish experiments
on sheet-wrapped plastic pipes tested on a site with heavy seepage of
ferrous-rich groundwater. So my question is to which extent your findings are verified in other Dutch experiments.

Answer of Mr. Ven

We concluded that the entrance resistance could not be reduced by
flushing from experiments on 4 or 5 pilot areas in the IJsselmeerpolders and in the Lauwerszee area. Other Dutch experiments with positive
results are not known to me. So I look forward to publications from
Denmark about this matter.

Questions from Mr. J. Wesseling to Mr. B. Tischbein

- Is chemical maintenance applied because of reasons of costs in Germany?
- Is chemical maintenance allowed?

Answer of Mr. Tischbein

In the Federal Republic of Germany maintenance is generally mechanical.
Chemical maintenance is only used exceptionally e.g. to complete the
mechanical method and in cases of unforeseen weed growth. As compared
to mechanical maintenance, chemical maintenance has the following disadvantages:

- applied alone it is not sufficient for complete maintenance (no removing of weeds, no clearing away of sediments);
- it may entail negative side effects:
  . direct toxic effects (very important if drainage water is used for
    drinking or irrigation)
  . weeds may get resistant to applied chemicals; problems of secondary
    flora are arising (change of chemicals, higher costs)
  . there is a risk of damage to the sidewalls of the watercourses by
    erosion (the stabilizing effect of plant roots may be vanished or
    reduced)
  . oxygen content of the water may decrease (disintegrating processes
    related to killed organic material)

Taking into account ecological aspects, the use of chemicals may be
very problematic (depending on the special situation and the way of
handling with chemicals). Economically chemical maintenance is more
expensive than the mechanical methods.

Question from Mr. G. Spoor to Mr. L.C.P.M. Stuyt, Mr. W. Diericks, Mr. G.A.
Ven

Progress using current empirical approaches for determining envelope
requirements for particular soil conditions seems to be levelling off
without the problem being solved. What future approach(es) would the
researchers suggest should be followed to provide the necessary design
information and how can cooperation between different workers be best
achieved?

Answer of Mr. Dierickx

I thought that we all three emphasized that cooperation is necessary.
There is already a cooperation in research on envelope materials and
related problems between the Dutch, French and Belgian investigators.
We are aware of what is going on in each country.

Answer of Mr. Stuyt

The problem is fundamental and difficult to solve. Basically, we still
do not understand the functioning of an envelope/soil system. We are
still in the empirical stage, in the sense that we monitor the functioning of an envelope/soil combination and we try to explain it. At
the Institute for Land and Water Management Research (ICW) we are going to measure the spatial variability in the envelope's hydraulic
conductivity as a function of the hydraulic gradient. We work at im-

218

proving the definition of a relevant pore size distribution. Two years of data from two pilot areas indicate the usefulness of sheet envelopes in the Netherlands under certain conditions: this is not in accordance with the current opinion in this country as regards the use of sheet materials. Empirical research is going on more in depth, and at the moment this seems to me the only way to make progress.

Answer of Mr. Dierickx

I have regular contacts with Mr. Stuyt and the French; this is beneficial for making progress in this field.

Question from Mr. B. Lesaffre to the whole panel

Mr. Ven proposes to use voluminous coarse envelopes in order to limit iron clogging. In France, we have some experience on the use of gravel backfills or substitutes. In some cases it was noticed that iron settles inside the permeable backfill instead of in the drain pipe itself, so no clogging occurs. But gravel is expensive and the ecological concern about its digging will likely increase. Has anyone of you any experience of using industrial waste (e.g. high temperature deformed plastic) that could be a cheap gravel substitute?

Answer

None of the panel members has any experience in this field.

Mr. Lesaffre

Because there is no answer, research on such materials might be necessary.

Session 4
Regional and local water management systems

Chairman: J.WESSELING
Institute for Land & Water Management Research (ICW),
Wageningen, Netherlands

Models for the assessment of sedimentation in reservoirs as a tool for correct management and lifetime estimate in the project phase

P.BAZZOFFI
Ministero Agricoltura e Foreste, Firenze, Italy

ABSTRACT

The construction of reservoirs, started in the fifties, has recently increased in Italy in view of the ever growing need for water for irrigation purposes. This increase is also caused by the publicity and financial support of the Ministry of Agriculture and Forestry, especially for hilly reservoirs.

The value for agriculture of such water storage reservoirs has been proved. Unfortunately severe siltation has occurred in a few cases, with considerable damage for farm and community economies. The siltation is mainly due to a lack of methods and to the inadequacy of existing models for sediment yield predictions in the project phase.

Through a field survey of sedimentation in three small reservoirs in central Italy, data were obtained and existing provisional models were tested. Various other data on sedimentation in 9 reservoirs all over Italy were collected from literature. Through the use of multiple regression techniques, sediment deposition was expressed as a function of measured reservoir and watershed variables.

The scope of the research was developing a model for sedimentation prediction, based on morpho-physiographic and climatic characteristics which are very easy to be obtained by planners. Although more observations are considered to be needed, the model obtained may be considered applicable over a wide range of conditions, and being a first effective tool in the project phase.

1. INTRODUCTION

In recent years in Italy there has been a marked growth in the construction of reservoirs for irrigation purposes. The increase has been made possible both by the availability of machinery which has made the movement of large volumes of soil easier, and by the financial support furnished by public corporations for realization of these works. The greatest increase occurred in the number of small and medium size reservoirs for farm use only. At present, there are more than 8400 reservoirs with a total storage capacity of more than 300 million cubic meters and an average capacity between 30 000 and 40 000 cubic meters (Medici, 1980). In the years 1960-1970 there was an increase of more than 400% in the number of reservoirs in the north, and of about 100% of those in central and southern Italy.

The topographyc characteristics of the national territory are such as to permit a further increase in the number of reservoirs, especially for the purpose of providing irrigation water in hill lands. In fact, hills cover about 12.7 million hectares, corresponding to approximately 40% of the Italian territory. At least 10-20% of these terrains are suitable for irrigation (Crivellari, 1938). From a practical point of view it is possible to increase the number of reservoirs existing today about ten times, which would mean about 85 000 units (Giorgi, 1965).

If in the near future it is intended to give a new impetus to the realization of such works, care should be taken not to repeat past errors. In some cases the lack of attention in the project phase has caused two types of faults, often occurring i.e. partial water-filling of the reservoir and a rapid siltation. This drawback, together with other causes, has caused a drastic decrease in the number of new reservoirs realized recently.

By means of the present study it is intended to achieve a three-fold aim:

1) to present a rapid and practical method for the direct measurement of sediments in a reservoir;
2) to evaluate existing sediment prediction models by means of a comparison between the results of these models and the actually measured sediments;
3) to develop a prediction model suitable for the Italian territory from the sedimentological and morpho-physiographic analysis of watershed reservoir systems of different sizes under a wide range of conditions.

The conceptual originality of the new model as compared to the existing ones is that its goal is to arrive at an estimate of the siltation on the basis of certain information of morphometric and physiographic nature, easily obtainable from the 'Official map of the State' (IGM) on a scale 1:25 000, and simple meteorological data. The maps 1:25 000 of the entire national territory have been published by the Military Geographic Institute. Meteorological data on a 30-year scale necessary for the investigation have been published by the Ministry of Public Works.

The possibility of using direct sediment measurement and of utilizing them for the above mentioned purposes is based on the fact that, in the Italian hill territory, there exists quite a number of artificial watershed reservoirs. By dividing the number of existing lakes by the total hill area, we obtain the average value of one lake every 14.3 km^2. Furthermore, we must consider that for some regions the number of reservoirs is higher than the average national value, e.g. in Emilia Romagna where, on an average, the lakes are at a mutual distance of only 1.7 km. In the regions with the smallest number of lakes, such as Puglia and Campania, the average distance between reservoirs is still very short, namely in the order of about 16 km. These estimates refer only to reservoirs built with a contribution of the State, so that the actual number of reservoirs must be considered greater, although it is not possible to give a precise estimate of them.

The presence of such a large number of reservoirs makes it possible for the planner to use direct measurements of the sediment, testing the forecasts obtained by means of estimation models. Obviously, such a comparison is possible under the condition that the morpho-physiographic characteristics of the watersheds are comparable. Meteorological conditions may be considered equal within the same region if distances between reservoirs are short and they have the same elevation.

2. EXPERIMENTAL PROCEDURE

In the forecast model siltation was considered as a variable dependent on a number of other quantities, according to the linear model of multiple regression. Through analysis of the data, we searched for the model that gave the best fitting and also the one that linked maximum significance of the overall equation and coefficients with the best fitting. Data of 12 watershed-reservoir systems of different size located in several parts of the

Fig. 1   Reservoir locations in Italy

225

national territory (Figure 1) were utilized. Direct measurement of sediment material was realized only for 3 small reservoirs located in Tuscany (Bazzoffi and Panicucci, 1983; Bazzoffi and Panicucci, in press) while for the remaining ones sedimentation data were utilized from investigations carried out for different purposes in the past by other authors (Visentini, 1939; Guggino, 1961; Rosa, 1964).

The effectiveness of the forecast models of Woodburn (1955), Gottschalk (1964) and Roehl (1962) was tested by comparing the values computed with their models and those effectively measured in the 3 reservoirs. Because Woodburn's formulas presupposes the knowledge of the gross erosion in the watershed and this information was not available for the other locations, it was not possible to extend the comparison to the remaining watersheds. Roehl's formulas were only applied to the three cases, because only for them all required data were available. The model proposed by Gottschalk and based on data from 18 small reservoirs in South Dakota, was tested for the 3 reservoir watersheds because this forecast model is valid only for smaller watersheds.

3. METHOD FOR DIRECT MEASUREMENT OF SEDIMENTATION

In order to measure the sediments of a reservoir different methods can be followed depending on factors like the dimensions of the reservoir, the accessibility of the shores, the necessity of repeating the measurement after a lapse of time, and the type of mathematical method for determining the volume of the silted material.

The method used here was the one developed for small reservoirs by Bazzoffi and Panicucci (1983). It substantially differs from the method used in the USA, which is more suitable for reservoirs of larger dimensions (USDA, 1979), and from other survey methods applied in Italy, e.g. Stoppini and Rossi (1978), Stoppini et al. (1979), Sensidoni and Mazzetti (1934).

The procedure consists of taking a certain number of transects along which measurements of the depth of the water and of the thickness of the sediment are performed. After a preliminary survey with an echograph and a simple acoustic metercounter (Bazzoffi and Panicucci, 1983), the measurements are carried out aboard an easily-assembled raft. To mark the transect, metal stakes are placed on the shores of the reservoir and between them a rope is spanned. This rope is used as a guide both for the metercounter (in the preliminary survey phase with the echograph) and for the raft which is moved along the section by the operators who are aboard.

Different from the USDA procedure, which applies 'prismoidal' formula, measurement along sections is not strictly essential; in fact, calculation of the volumes of water and sediment can be made from a scattered network of points.

For the bathymetric measurements, a metric type-measure furnished with a metal disk with a dimension appropriate to indicate the sediment surface is utilized. The thickness of the sediment (in the same point in which the bathymetric measurement is made) is determined by lowering a drill stick composed of a series of tubes which can be screwed on to each other. The point of the drill is equipped with a small metal disk so it it easy to perceive the reaching of the primitive bottom. The sediment thickness is obtained from the difference between the two measurements.

For each measuring point, the position of the raft is surveyed in order to enable computerized analysis of data for determining the volumes of water and sediments.

The density of the sediments is determined by means of core borings made with a special sphere-valve drill corer, esspecially constructed for core boring in soft sediments (Bazzoffi, 1981). This device permits both a minimum disturbance of the sediment and the possibility of breaking the core down into sections, so the density along the entire sedimentary profile is obtained.

The value of the average annual net erosion is obtained by dividing the total dry weight of sediment by the surface area of the watershed and by the number of years of operation of the reservoir.

4. EVALUATION OF EXISTING MODELS

Evaluation of forecasting models was made for obtaining an indication which of them would predict the sediment value closest to that actually measured and to select the most significant independent variables to be introduced in the new model.

The estimate of gross erosion with the Woodburn's formulas requires a parameter which is always difficult to determine. Generally, the prediction models which take into consideration this variable use the equation of Musgrave (1947) or the Universal Soil-Loss equation of Wischmeier and Smith (1965). These equations are statistical models commonly used in the USA for computing the rill and inter-rill erosion at field scale, and are based on observations obtained from small experimental plots. However, the authors of these models warn against applying these instruments to the hy-

drographic watersheds where other forms of erosion can play a role, together
with the surface erosion caused by water.

In a previous study conducted on the 3 aforementioned watersheds
(Bazzoffi, 1985), gross erosion was determined also by means of other procedures for the application of quantitative models of Gavrilovic (1959) and
of Zemlijc (1971) for estimating the net production of sediment by hydrographic watersheds. The values of gross erosion estimated with the three
models USLE, GAVRILOVIC and ZEMLIJC were therefore utilized as independent
variables in the models to be tested.

In the application of the Gabrilovich and Zemlijc models it was found
that it was necessary to carry out separate calculations for wooded areas
and for cultivated areas, and then to summarize the separate values obtained. For several factors data was available, including minima and maxima, so
that the results covered a wide range of data. These data, utilized in the
models, have furnished the results reported in Table 2.

The models applied, expressed in metric units, are as follows (Matarrese,
1978):

$$St = \frac{A^{0.8957} Sg^{0.8573} Cs^{0.3423} Tn^{0.6573}}{4.057} \qquad \text{(Woodburn)} \qquad (1)$$

$$St = 2.492 \ A^{0.9151} \ Sg^{0.8303} \ Tn^{0.7329} \qquad \text{(Woodburn)} \qquad (2)$$

$$S\% = 2.82687 - 0.26 \ \log 10A - 0.531 \ \text{colog ia} \qquad \text{(Roehl)} \qquad (3)$$

$$S\% = 4.59574 - 0.23043 \ \log 10A - 0.51022 \ \text{colog ia} - 2.78594 \ BR \quad \text{(Roehl)} \quad (4)$$

$$Sm = 0.0522 \ C + 8.22 \ A + 330.68 \ Tn - 2216.95 \qquad \text{(Gottschalk)} \qquad (5)$$

TABLE 1   Factors for the validation of the existing models

Factors	Reservoir-watershed systems		
	Cavalcanti	Pavone	Mandracco
A	60.37	54.54	75.91
Cs	3035	1907	2853
Tn	25	25	19
C	183247	104044	216615
ia	0.03889	0.03426	0.01416
BR	4.64	4.42	3.73
SgGAVR.	337.21/549.43	211.37/344.61	185.98/303.12
SgZEML.	119.22/194.25	74.39/121.87	69.07/107.14
SgUSLE	5.3	2.9	0.9

TABLE 2   Comparison between observed and predicted values (t/ha/year)

Values	Cavalcanti			Pavone			Mandracco		
Observed	8.7			4.6			5.0		
Predicted (gross ero/n)	Gavr.	Zeml.	USLE	Gavr.	Zeml.	USLE	Gavr.	Zeml.	USLE
Models:									
Woodburn (1)	122/185	50/76	3.5	70/107	29/44	1.8	77/117	33/48	0.8
Woodburn (2)	94/140	39/59	3.0	64/96	27/41	1.8	60/90	27/38	0.7
Roehl (3)	76/124	27/44	1.2	29/47	10/17	0.4	23/38	9/13	0.1
Roehl (4)	81/131	29/47	1.3	56/91	20/32	0.8	46/75	17/27	0.2
Gottsch. (5)	17			14			18		

where St = ton of sediment

   S% = sediment as % of erosion

   Sm = m^3 of sediment

   A = watershed area in ha

   ia = R/L ratio (where R = average elevation of watershed perimeter with
        respect to the waterhsed outlet, L = average length of tribu-aries
        taries)

   BR = bifurcation ratio (Strahler, 1957)

   Sg = gross erosion in tons/ha per year

   Cs = ratio between reservation capacity and watershed surface (m^3/ha)

   Tn = age of the reservoir in years

The data necessary for the calculations are reported in Table 1.

From the tables it is clear that the sediment computed values are in all cases very different from the observed values.

5. SELECTION OF VARIABLES

The variables taken into consideration for the formulation of the new models were chosen by considering the more significant variables in the models tested, and by other factors considered to be of influence in the ablation-sedimentation process and mentioned by Strahler (1964), Wischmeier (1965), Ven te Chow (1964).

As mentioned above, the porphometric and physiographic data were obtained from topographical maps 1:25 000, while values for the climatic variables were taken from the 30-year data published by the Hydrographic Office of the Ministry of Public Works.

The chosen variables are the following:

TABLE 3   Regression analysis (eq. 6)

Analysis of variance

Source	Deg. of freedom	Mean square	F-value
total	11		
regr.	9	552600.93216	130.7526
W	1	3166758.633	749.2963
Es	1	77125.24953	18.2488
Rc	1	1175578.625	278.1572
Wp	1	12370.23966	2.9270
Rd	1	97235.57417	23.0072
Pr	1	133475.17372	31.5820
BR	1	65041.33792	15.3896
Sf	1	67218.72867	15.9048
LP	1	178604.45124	42.2601
Resid Resid	2	4226.31025	

R-squared   = .998
Standard error of EST.   =   65.010

Regression coefficients

Variable	Std. format	Std. error	T-value
Const.	1993.87693	565.54460	3.625
W	−26.96108	4.08738	−6.596
Es	12.914.60	4.38774	2.943
Rc	5.83029	1.21808	4.786
Wp	60.26760	10.99396	5.482
Rd	−27.15907	6.64042	−4.090
Pr	1.18090	.27467	4.299
BR	−91.88028	61.20335	−1.501
Sf	−35.39105	14.93020	−2.370
LP	−349.24182	53.72307	−6.501

Residual analysis

Observation number	Observed Y	Predicted Y	Residual	Standardized residual
1	1441.61000	1439.34557	2.26443	.03483
2	1492.64000	1506.09052	−13.45052	−.20690
3	2023.84000	2023.39904	.44096	.00678
4	1697.00000	1694.17895	2.82105	.04339
5	541.87000	492.77573	49.09427	.75518
6	288.05000	360.48066	−72.43066	−1.11415
7	309.53000	310.75419	−1.22419	−.01883
8	497.00000	489.76092	7.23908	.11135
9	300.00000	309.07665	−9.07665	−.13962
10	10.00000	5.36535	4.63465	.07129
11	550.00000	536.48493	13.51507	.20789
12	363.00000	346.82749	16.17251	.24877

Significance test:	source	d.f	SS	MS	F
	regr.	9	4976492.053	552943.56	205.98
	dev.	2	5368.960	2684.48	
	tot.	11	4981861.013		

K   = m^3 of dry sediment stored per km^2 of watershed surface

W   = watershed surface in km^2

Es   = erodable surface (ploughed lands + 1/16 non-ploughed lands such as woodland, rangelands, pastures, structures, etc.)

Rc   = reservoir capacity in $m^3.10^{-5}$

Wp   = watershed perimeter in km

BR   = bifurcation ratio (Strahler, 1957)

L1° = total length of first order channels in km

Lm1° = average length of first order channels in km

L   = total length of hydrographic network in km

Dr   = drainage density (Horton, 1932)

Sf   = stream frequency (Horton, 1932)

PmT = average slope gradient of principal(s) tributary(ies) in %

S   = average slope of watershed surface in %

Lo   = length of overland flow (Horton, 1945)

P   = $(0.43 + S + 0.0435^2)/6.613$ (Wischmeier, 1965; Arnoldus, 1977)

LP   = $(Lo/22.1)^m.P$, where m = 0.3 for s less than 0.5% and 0.6 if s is more than 10%

De   = difference between the average elevation of watershed perimeter and the average elevation of principal(s) tributary(ies)

Rd   = rainy days per year (average over a period of 30 years of observations

ggPFo = Fournier's index computed for rainy days (computed in the same way as RFo)

Pr   = average precipitation in mm per year (over a period of 30 years)

Rfo = rain erosivity index by Fournier (FAO, 1979) defined here as $\sum_{1}^{12} \frac{Pm^2}{P}$ where Pm = monthly mean precipitation in mm and P = mean annual precipitation in mm

ER   = elongation ratio (Shumm, 1956)

Rf   = form factor (Horton, 1932)

Cr   = circularity ratio (Miller, 1953)

mct = total cubic meters of sediment stored per year

y   = years between surveys

6. REGRESSION ANALYSIS

To determine the best combination of variables for predicting reservoir siltation, multiple linear regression analyses were performed. Data manipulation was run through an HP86B computer and its 'Regression analysis pac'

TABLE 4   Regression analysis (eq. 7)

Analysis of variance

Source	Freedom	Mean square	F-value
total	11		
regr.	7	709840.63067	218.8064
W	1	3166758.68	976.1447
Es	1	77125.24953	23.7736
Rc	1	1175578.96	362.3690
S	1	152941.04852	47.1437
Wp	1	195662.38182	60.3124
Rd	1	20498.17327	6.3185
Pr	1	180319.92698	55.5831
Resid	4	3244.14882	

R-squared   = . 997
Standard error of Est.   =   56.95743

Regression coefficients

Variable	Std. format	Std. error	T-value
Const.	1816.20078	271.47871	6.690
W	-23.06357	2.87073	-8.034
Es	12.23431	2.76624	4.423
Rc	6.68510	.79057	8.456
S	-15.98971	1.64547	-9.717
Wp	48.84193	7.11969	6.860
Rd	-27.16992	4.78502	-5.678
Pr	1.08965	.14616	7.455

Residual analysis

Observation number	Observed Y	Predicted Y	Residual	Standardized residual
1	1441.61000	1422.46648	19.14352	.33610
2	1492.64000	1432.50928	60.13072	1.05571
3	2023.84000	2017.00570	6.83430	.11999
4	1697.00000	1758.22184	-61.22184	-1.07487
5	541.87000	499.64671	42.22329	.74131
6	288.05000	297.57512	-9.52512	-.16723
7	309.53000	304.12635	5.40365	.09487
8	497.00000	547.44373	-50.44373	-.88564
9	300.00000	301.49819	-1.49819	.02630
10	10.00000	4.74195	5.25805	.09232
11	550.00000	575.32032	-25.32032	-.44455
12	363.00000	353.98432	9.01568	.15829

Significance test:	source	d.f.	SS	MS	F
	regr.	4	4976186.66	710883.81	501.12
	dev.	7	5674.35	1418.59	
	tot.	11	4981861.01		

232

developed by Boardman of the Statistical Laboratory of Colorado State University.

Using different sets of independent variables, selected from the listed factors, the regression equation that fitted data best was:

$$K = 1993.88 - 26.96W + 12.91Es + 5.83Rc + 60.27Wp - 27.16Rd + 1.18Pr -$$
$$- 91.88BR - 35.39Sf - 349.24LP \tag{6}$$

Table 3 shows the principal statistic parameters of the regression analyses. The hypothesis that the entire regression equation does not explain a significant part of the variation in K is rejected by the test of significance. Since the tabled value $F_{995,9-2}$ is 99.38 one can conclude that the overall equation is highly significant.

From the calculated T values, all the regression coefficients, with the exception of BR and Sf, are significant at 5% level since the tabled $t_{95,2}$rs 2.92 furthermore W, Rc, Wp, Pr, LP are significant at the 0.1% level. Coefficients of T associated to BR and Sf, are significant between 5 and 10% probability.

Another equation which associates the significance of all the coefficients with the best fitting was selected. By taking BR, Sf and S instead of LP, which is a difficult to determine parameter, the equation gets the form:

$$K = 1816.20 - 23.96W + 12.23Es + 6.69Rc - 15.99S + 48.84Wp - 27.17Rd +$$
$$+ 1.09Pr \tag{7}$$

The principal statistical parameters of this regression analysis are shown in Table 4. This equation also explains a significant variation in K as may be seen from the value of coefficient $F._{995.7-4}$ = 21.6. All the regression coefficients are highly significant at 0.05% level while Es is significant at 0.25% level; in fact tabled values are $t_{995.4}$ = 4.60 and $t_{975.4}$ = 2.78.

Although r^2 determination coefficients are very close to 1, the number of coefficient estimates exceeds 35% of the number of observations. To avoid the risk of 'over-fitting' it would be necessary to increase the number of observations at least to 30 and 23 respectively for the first and the latter model.

The range, mean and standard deviations of the independent variables considered in the models are as follows:

233

Factors	Units	Range values	Mean	St. Dev.
W	km^2	0.54 – 241.62	44.15	69.72
Es	km^2	0.11 – 186.40	26.56	53.75
Rc	$m^3 \cdot 10^{-5}$	1.04 – 140	43.67	57.92
Wp	km	2.8 – 80	23.88	22.77
Rd	–	65 – 118	87.50	16.42
Pr	mm	509 – 1800	986.91	379.10
BR	n.d.	3.47 – 4.64	4.13	0.52
Sf	–	1.19 – 18.76	6.48	5.65
LP	–	0.169 – 3.71	0.65	0.98
S	%	14.07	30.34	17.64
K	$m^3 \cdot km^{-2} \cdot a^{-1}$	10 – 2023.84	792.88	672.96

7. DISCUSSION AND CONCLUSIONS

From the statistical analysis it can be seen that both equations are good predictors of the amount per year and per km^2 of sediment stored in the reservoir. As a first step in this research, eq. (7) seems to be preferred, both because the values of the parameters are easier to determine, and also because the high value of r^2 is a high significance of the entire equation and of its coefficients (see Figure 2).

As all forecasting models with a statistic basis, also these two models should not be applied to reservoir-watershed systems having values for independent variables greater or less than the ones used here.

In order to apply these models in the project phase, it must be realized that the principal limitation of these equations is the limited number of observations they are based upon. However, compared to the existing models, they appear to be slightly better as a prediction tool.

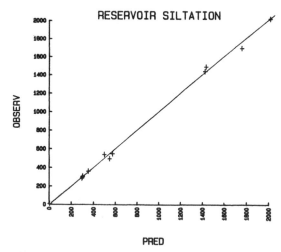

Fig. 2   Relationship between predicted and measured values of the dependent variable in eq. (7)

The very scattered location of the reservoirs over the territory and the very different environmental characteristics could partly balance the limitations introduced by the risk of overfitting. Through an analysis of the regression coefficients and their sign, it is possible to make a theoretical comparison between the role attributed to them in the equation and that in using them for forecasting.

The most important factors in determining siltation proved to be reservoir capacity and watershed surface. This is in agreement with the theoretical formulation of the sediment process. In fact, the greater the capacity of the reservoir, the more opportunity is given for siltation as already noted by Dendy (1966) (see the plus sign for the coefficient).

On the contrary, the larger the watershed surface, the higher the internal re-deposit of eroded materials (Maner and Barnes, 1953; Gottschalk and Brune, 1950; Roehl, 1962) which corresponds to a minus sign for the coefficient.

The erodible surface, Es, is directly related to sedimentation and this is expressed by a positive value of the coefficient.

The perception that the two surfaces (the total surface and the erodible surface) should constitute two independent variables has turned out to be very important. In fact, by considering only the total surface, a considerable residual variance would have been obtained.

Another factor which plays a very important role is the average slope of the watershed. When this variable is small, it could act in favour of soil erosion, while when it increases the variability explained must likely be referred to the fact that generally on steeper slopes very erodible soils are rare to find, or if they are present they are generally covered by forest or rangelands. The negative value of the coefficient may agree with prevailing conditions of the latter type in the set of observations. From this consideration it could be profitable to split this variable into two new variables: 1) average slope of ploughed lands and 2) average slope of non-ploughed lands.

The perimeter of a watershed is directly related to siltation, probably because longer perimeters correspond with lower peaks of deflux which influences the sedimentation process positively (Horton, 1964). Soil erosion is related to rain erosivity which is directly dependent on the amount and intensity of rain. This variability is well explained by the positive coefficient associated to the average amount of rain per year and by the negative value of the coefficient of rainy days which expresses in part the

235

concentration of the rain.

For planning purposes, it is important to make clear how extension of the ploughed surface plays a role in siltation of the reservoirs. For this reason, it will be useful to indicate already in the project phase the surfaces to be maintained as woodland or to be reafforested in order to avoid any acceleration of the erosion-sedimentation process.

The research carried out on the 12 watershed-reservoir systems appears to be very successful, both in terms of enlarging our knowledge of the phenomena which govern the sedimentary processes, as well as for the immediate practical applications of the results in the field of engineering when dealing with water resources for agricultural, civil and industrial uses.

The results obtained clearly indicate the need for continuing the research in order to increase the reliability of the models.

ACKNOWLEDGMENTS

The author wishes to thank Prof. G. Chisci for this encouragement and the useful suggestions given by him.

REFERENCES

Arnoldus, H.M.J. 1977. Predicting soil losses due to sheet and rill erosion. FAO Conservation Guide 1: 99-124.
Bazzoffi, P. 1981. Sphere type corer for sampling soft sediments in reservoirs. Annali Ist. Sper. Studio e Dif. Suolo, Firenze. Vol. XII.
Bazzoffi, P. and Panicucci, M. 1983. Erosione sui versanti e conseguente sedimentazione in piccoli serbatoi artificiali. Nota I. Confronto fra il valore risultante dai rilievi desimentometrici e dal bilancio del radionuclide Cs-137 da fallout derivato da esplosioni nucleari. Annali Istituto Sperimentale per lo Studio e la Difesa del Suolo, Firenze. Vol. XIV.
Bazzoffi, P. and Panicucci, M. (in press). Erosione sui versanti e conseguente sedimentazione in piccoli serbatoi artificiali. Nota II. Risultati sperimentali per i bacini 'Pavone' (Valdera) e 'Mandracco' (Mugello).
Bazzoffi, P. (in press). Erosione sui versanti e conseguente sedimentazione in piccoli serbatoi artificiali. Nota II. Applicazione della equazione universale per le perdite di suolo a tre bacini Toscani.
Crivellari, G. 1983. Laghetti collinari. Edagricole Bologna. 272 pp.
FAO. 1979. A provisional methodology for soil degradation assessment. Rome.
Dendy, F.E. 1974. Sediment trap efficiency of small reservoirs. ASAE Transactions 17(4): 898-901, 908.
Gavrilovic, 1959. Méthode de la classification des bassins torrentielles et équations nouvelles pour le calcul des hautes eaux et du débit solide. Belgrado (in Jougoslav). (manuscript)
Giorgi, E. 1965. I laghetti collinari in Italia. Notizie sulla diffusione e sul loro inserimento nell'economia dell'azienda collinare. I. NEA Oss. per la Toscana, Firenze.
Gottschalk, L.C., Brune and Gunner, M. 1950. Sediment design criteria for

the Missouri basin loess hills. Soil Conserv. Serv. USDA (Techn. Publ.) SCS-TP-97. Milwaukee, Wisconsin.

Gottschalk, L.G. 1964. Reservoir sedimentation. In 'Handbook of applied hydrology' (ed. Ven Te Chow). McGraw-Hill, New York.

Guggino, Picone, E. 1961. Sui contributi di piena da piccoli bacini. Ist. Idr. Agr., Catania: 47-80.

Horton, R.E. 1932. Drainage basin characteristics. Trans. Am. Geophys. Union 13: 350-361.

Horton, R.E. 1945. Erosional development of streams and their drainage basins; hydrophysical approach to quantitative morphology. Bull. Geol. Soc. Am. 56: 275-370.

Maner, S. 1958. Factors aspecting sediment delivery rates in the Red Hills physiographic area. Trans. AGU 39.4: 669-675.

Medici, G. 1980. L'irrigazione in Italia, dati e commenti, ed. Agricole, Bologna.

Miller, V.C. 1953. A quantitative geomorphic study of drainage basin characteristics in the Clinch Mountain area, Virginia and Tennessee, Project nr. 389-042. Techn. Rep. 3, Columbia University, Dept. of Geology, ONR, Geography Branch, New York.

Musgrave, G.W. 1947. Quantitative evaluation of factors in water erosion, a first approximation. Journ. of Soil and Water Cons. 2.3: 133-138.

Piccoli. 1951. L'interrimento del serbatoio di Molato (Tidone) nel periodo 1925-1949. L'acqua, fascicolo, 7-8.

Roehl, J.W. 1962. Sediment source areas, delivery ratios and influencing morphologica factors. Ass. Int. Hydr. Scient. Publ. 59, Gentbrugge: 202-213.

Rosa, G. 1964. Prima indagine sull'interrimento dei laghetti collinari in Sardegna. Estr. da Studi Sassaresi, sez. III. Annali della Facoltà di Agraria dell'Univ. Sassari. Vol. XI. 26 pp.

Schumm, S.A. 1956. Evolution of drainage systems and slopes in badlands at Perth Amboy, New Jersey. Bull. Geol. Soc. Am. 67: 597-646.

Sensidoni, F. and Mazzetti, A. 1934. Il trasporto solido nei corsi d'acqua italiani. Pub. 15 Serv. Idr. J.LL.PP.

Stoppini, Z. and Rossi, R. 1978. Indagine sperimentale sull'interrimento di alcuni laghetti collinari e sulla erosione dei versanti. I. Metodologia di rilievo della sedimentazione. CNR, Pr. Fin. Cons. Suolo, pub. 24.

Stoppini, Z., Rondelli, F. and Paxoletti, M.T. 1979. Indagine sperimentale sull'interrimento di alcuni laghetti collinari e sulla erosione dei versanti. II. Misura del volume e del peso del sedimento totale prodottosi in due laghetti della Valtopina (Umbria). CNR, Prog. Fin. Cons. Suolo, pub. 57.

Strahler, A.N. 1957. Quantitative analysis of watershed geomorphology. Trans. Am. Geophys. Union 38: 913-920.

Strahler, A.N. 1964. Quantitative geomorphology of drainage basins and channel networks. In 'Handbook of applied hydrology' (ed. Ven Te Chow). McGraw-Hill, New York.

USDA. 1979. Field manual for research in agricultural hydrology. USDA Handbook 224.

Ven Te Chow. 1964. Handbook of applied hydrology. McGraw-Hill Company, New York. 1418 pp.

Visentini, G. 1939. Depositi alluvionali nei serbatoi italiani e trasporto solido fluviale. L'energia elettrica.

Wischmeier, W.H. and Smith, D.D. 1965. Predicting rainfall-erosion losses from cropland east of the Rocky Mountains. Guide for selection of practices for soil and water conservation. Agric. Handbook 282.

Wischmeier, W.H. and Smith, D.D. 1978. Predicting rainfall-erosion losses. A guide to conservation planning. USDA Handbook 537.

Hydrological and economical effects of manipulating water levels in open water conduits – A case study

P.J.T.VAN BAKEL
Institute for Land & Water Management Research (ICW), Wageningen, Netherlands

ABSTRACT

Drainage-boards and polder districts in the Netherlands try to fulfil the agricultural demands with respect to water management by manipulating water levels in open water conduits. To get a more scientific basis for optimal manipulation, a case study has been carried out. In the framework of this study a model has been developed to simulate different alternatives of manipulating surface water levels. In this way the hydrological effects of maintaining high surface water levels during summer and of water supply for subsurface irrigation are obtained. By converting the hydrological effects in additional crop production, benefits and costs, the internal rate of return of projects for water conservation and water supply are calculated.

1. INTRODUCTION

In the Netherlands the demand for water by agriculture, industry and households is steadily increasing. This calls for proper water management schemes at all management levels, i.e. national, provincial, drainage-board or polder district and users level. This paper deals with the water management at drainage-board level.

In the Netherlands a drainage-board or polder district is a public organization, supervised by the provincial government, that takes care of the surface water management in a certain area. The main task of drainage-boards is maintenance of the main drainage system and the manipulation of the water level in the open conduits.

The seasonal trend in weather conditions causes a seasonal trend in the depths of the groundwater table with high levels in winter and low ones during summer while from agricultural point of view this trend should be the opposite. Drainage-boards in the Netherlands try to fulfil the agricultural demands by manipulating water levels in open water conduits. During winter these water levels are kept as low as possible, while in spring they are raised to conserve water in the area. In case external water supply is possible, the water level can be kept high during the whole growing season so that subsurface irrigation (sub-irrigation) can be realized.

Because weather conditions vary considerably from year to year and also local hydrological conditions are different, no standard procedure for manipulating surface water levels can be given. In fact, the decision upon the most desired level at any time is an optimization problem. In practice, however, the manipulation is done by rough thumb rules.

To get a more scientific basis for the manipulating of surface water levels, a case study has been carried out in a pilot area of 8000 ha in the northeastern part of the Netherlands. This paper gives a short description of this area and the developed model to simulate effects of manipulating of surface water levels on crop transpiration. Next the conversion of hydrological effects to additional crop production, costs and benefits will be treated.

2. DESCRIPTION OF THE PILOT AREA

2.1. General

The study area is part of a reclaimed cut-over raised bog district (Figure 1). The altitude varies between 7 and 10 m above Ordnance Datum. It covers approximately one third of the territory of the drainage-district 'De Veenmarken'.

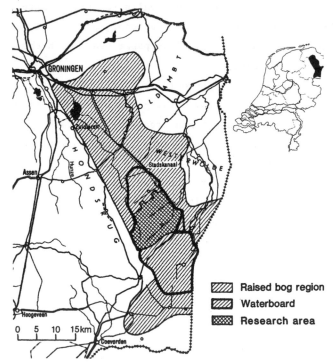

Fig. 1   Geographical position of the research area 'De Monden' and the drainage-district 'De Veenmarken'

The type of agriculture is mainly arable farming and the crop rotation scheme consists of approximately 50% potatoes for starch production, 25% sugar beets and 25% cereals. Agriculture constitutes the only interest in surface water management in the region.

2.2. Geo-hydrological and soil-physical properties

The hydrological basis of the saturated groundwater system is formed by clayey deposits of tertiary age at a depth of about 120 m below soil surface. The aquifer above this base consists of 40 metres of mostly coarse sand deposited by rivers. During the Cromerian Interglacial a layer of clay and peat was deposited which is still (partly) present at some places. During the Saalian Glacial an ice-pushed ridge called the 'Hondsrug' has been formed. This ridge forms the western border of the area. The melting water eroded a wide and deep valley east of this ridge, which was later on filled with rather coarse sand. These deposits form a second middle-deep aquifer with a thickness of approximately 25 m. During the Weichselian Glacial wind erosion deposited cover sands of approximately 15 m thickness. This layer forms the phreatic aquifer.

The relatively warm and wet climate during the Holocene has led to the formation of a vast raised bog area east of the 'Hondsrug'. Since the beginning of the 17th century peat was harvested in a systematic way by digging a system of canals which were used for the shipping of the dried peat.

During the peat harvesting, the upper 0.5 m of the peat, left behind because it was not suitable for burning, was levelled and covered with sand from the excavated canals. In this way a typical soil profile was created with a large water holding capacity, but poor fertility. The arable land use caused a steadily decomposition of the peat layer and the water holding capacity decreased gradually. To stop this process and also to improve the vertical water movement hampered by locally present impermeable layers at the boundary between peat and subsoil, soil improvement (sub-soiling) has been carried out on a large scale. This process is still going on.

For modeling the hydrological system of the area, the properties of the saturated, unsaturated and surface water system had to be known.

The spatial distribution of transmissivities of aquifers and c-values of flow resisting layers has been derived from already existing data, additional pumping tests and calibration with the finite element model FEMSAT (Querner, 1984; Smidt, 1984).

The spatial distribution of soil-physical properties has been derived by combining the soil mapping units of the available soil map 1:50 000 in eight soil-physical units, each unit having a particular sequence of soil layers. From each soil layer the soil water retention curve and the hydraulic conductivity as a function of the water content have been determined (Boels et al., 1978; Bloemen, 1980a,b).

2.3. The surface water system

The system of main and secondary canals, excavated during the peat harvesting, was initially used for shipping agricultural products and for drainage. The 150 to 200 m distance between the secondary canals was too large for a sufficient drainage, because shipping demanded high water levels. Therefore a main ditch midway between the canals was dug. In the time (about 1960) road transport became dominant. This resulted in lowering of the water levels for drainage purposes and the filling of many of the secondary canals and the main ditches. The remaining main canals and part of the secondary canals came under supervision of the drainage district. In fact the original secondary canals now form the main drainage system.

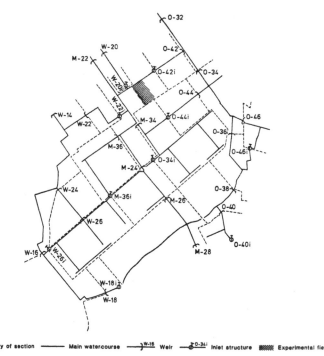

Fig. 2   Water management plan 'De Monden' and division of the area in subregions with the same surface water level

From an agricultural point of view the ideal seasonal level in the open watercourses is low during winter. In spring the water level is raised to conserve water. In case external water supply is possible, the surface water level can be kept high during the whole growing season and subsurface irrigation can be realized.

To be able to manipulate the surface water level in the main drainage courses the drainage district divided the area into sub-areas with different open water levels maintained by automatic weirs. Together with the construction of the weirs a number of inlet structures has been built (Figure 2).

The interaction between the surface water system and the groundwater system in principle can be derived from the geometry of the watercourses and the aquifer properties (Ernst, 1962). The mathematical form gives:

$$T = \frac{h_m - h_o}{v_f} \quad (d) \tag{1}$$

where T represents the so-called drainage resistance, h_m is the height of the phreatic surface midway between two parallel ditches (m), h_o is the open water level (m) and v_f is vertical flux density through the phreatic surface.

Use of eq. (1) is very attractive for modeling because of the linear one-to-one relation between h_f, h_o and v_f. From field measurement it has been derived that this form is valid not only for drainage but can also be used under sub-irrigation conditions. By simultaneously measuring the discharge from a particular area and the average value of h_f-h_o in that area, the average value of T can be found. In this way the drainage resistance for each sub-area has been derived (Smidt, 1984).

3. SIMULATION OF EFFECTS OF MANIPULATING SURFACE WATER LEVELS

3.1. Modeling the hydrological system

To find the effects of manipulating surface water levels one can either use field experiments or hydrological modeling. Field experiments are not only very time consuming, but hold only for the circumstances under which they are carried out. Therefore hydrological modeling is preferred.

The model perception is given in Forrester notation in Figure 3A (for the sub-irrigation situation) and Figure 3B (for the drainage situation). According to these diagrams the following hydrological sub-systems can be distinguished:

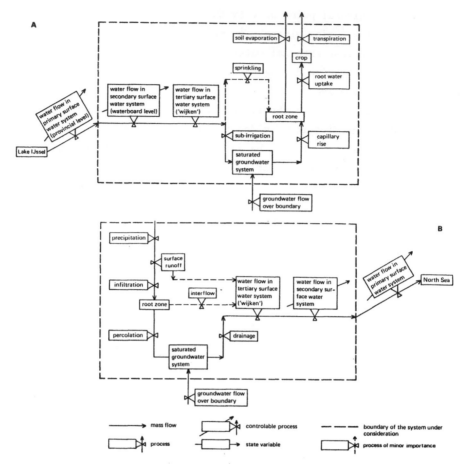

Fig. 3   Modeling of the hydrological system involved in the process of
sub-irrigation (A) and drainage (B)

- atmosphere - crop system
- unsaturated groundwater system
- saturated groundwater system
- surface water system

A regional 3-dimensional model in which all sub-systems are incorporated
would result in a too complex model. For simulating the effects of surface
water level manipulation on groundwater depths and crop transpiration therefore a one-dimensional model, called SWAMP (Surface WAter Management Program) has been developed. It consists of the following sub-systems (Figure
4):

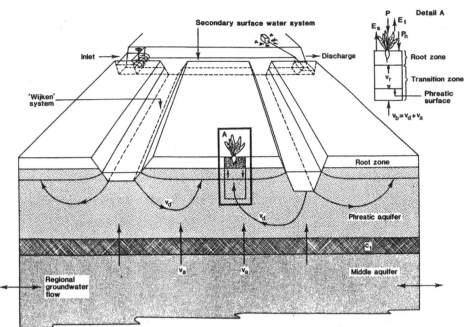

Fig. 4   Schematic representation of the hydrological system modeled
with the SWAMP-model

a) The unsaturated/saturated system above the lowest possible phreatic surface, in which water flow is vertical only

This sub-system is divided into a root zone and a transition zone. The
upper boundary conditions for the root zone are formed by transpiration
and evaporation. These have been calculated with the SWATRE-model
(Belmans et al., 1983), using standard meteorological data, known soil-
physical data, and data on crop height and soil cover. The lower boundary condition is given by capillary rise or percolation. These fluxes
are calculated from groundwater depth and water content in the root zone.
From the water balance of the root zone the course in time of the water
content in the root zone results. Because reduction in potential transpiration is modeled as a function of this amount of water, also actual
transpiration can be simulated.

The upper boundary condition of the transition zone is the capillary
rise towards or percolation from the root zone. The lower boundary condition of the transition zone is the volume flux density through this
boundary, v_b, being (Figure 4):

$$v_b(t) = v_a(t) + v_d(t) \qquad (m \cdot d^{-1}) \qquad (2)$$

245

where v_a is volume flux density to or from the deeper aquifers $(m \cdot d^{-1})$ and v_d is volume flux density to or from the surface water system $(m \cdot d^{-1})$. According to the quasi-stationary approach (De Laat, 1980) the storage in the transition zone depends on both groundwater depth and volume flux density of capillary rise or percolation. From the change of water content in the transition zone the change of the groundwater depth is calculated.

b) Saturated groundwater system

This system is described by the functional relationship of eq. (2). The groundwater flow to or from the surface water system can be written as:

$$v_d(t) = \frac{\overline{h_g^*}(t) - h_{o,t}^*(t)}{\eta \cdot T(h_f^*)} \tag{3}$$

where $\overline{h_g^*}(t)$ is average groundwater depth below soil surface, $h_{o,t}^*$ is depth of surface water, η is shape factor (-) and T is drainage resistance.

With the groundwater flow to or from the deep aquifer, v_d, the regional pattern of seepage can be incorporated. From calculations with the finite element model for stationary saturated groundwater flow FEMSATS this regional pattern has been calculated as a function of $\overline{h_f^*}$ and h_o^*, hence:

$$v_a(t) = f\{x, y, \overline{h_f^*}(t), h_o^*(t)\} \tag{4}$$

c) Surface water system

This sub-system is divided into a secondary and a tertiary surface water system. The tertiary system is considered as a reservoir. The state of the reservoir is the surface water depth, $h_{o,t}^*$, to be calculated from its water balance:

$$h_{o,t}^*(t + \Delta t) = h_{o,t}^*(t) + \Delta t \{v_d(t) + v_{o,s}(t)\}/a_t \tag{5}$$

in which $v_{o,s}$ is volume flux density to or from the secondary system $(m \cdot d^{-1})$ depending on the difference in surface water level between tertiary and secondary system. Because this system is incorporated in a one-dimensional model, its relative area, a_t, has to be taken.

The secondary system is also considered as a reservoir, with the water balance:

$$h^*_{o,s}(t + \Delta t) = h^*_{o,s}(t) + \Delta t\{-v_{o,s}(t) + v_{o,w}(t) - v_{o,p}(t)\}/a_s \qquad (6)$$

where $h^*_{o,s}$ is surface water depth, $v_{o,w}$ is flux density over the weir, $v_{o,p}$ is supply flux density from inlet structure and a_s is relative area of the secondary system.

d) Adjustable weir

The model image of this structure is a level (the weir crest) which can be moved up- or downwards. The flux density over the weir, $v_{o,w}$, is calculated from the head-discharge curve of the weir.
The actual manipulation of the surface water level is done through manipulating the weir and the supply rate at the inlet structure. Whether or not the weir is lowered or raised depends on the most desired surface water level at any time, the so-called target level. No standard rules can be given for the most favourable target level. In the model (and also in practice) the target level is governed by a number of operational rules.

The effects of manipulating the surface water level on crop transpiration are calculated using the chain: depth of weir crest or supply rate → water depth in the secondary surface water system → water depth in the tertiary surface water system → drainage flux → groundwater depth → capillary rise → amount of water in the root zone → transpiration. To obtain a solution an interactive system is used with an iterative numerical procedure and applying time steps of one day.

Verification of the SWAMP-model has been attempted with lysimeter data collected in the area during 1980 and 1981. Unfortunately, during these particular years no reduction in transpiration due to drought occurred. Fortunately, in 1982 there was a significant reduction in transpiration. The regional pattern of reduction in transpiration on a particular date could be derived from thermal infrared images. The reduction in transpiration from this pattern has been compared with the reduction in transpiration calculated with the SWAMP-model for the same places and the same time. In general, there was a good agreement between recorded and simulated reductions. For detailed information, the reader is referred to Nieuwenhuis et al. (1985).

3.2. Operational rules for the day-to-day surface water manipulation

The establishment of the best operational rules is in principle an op-

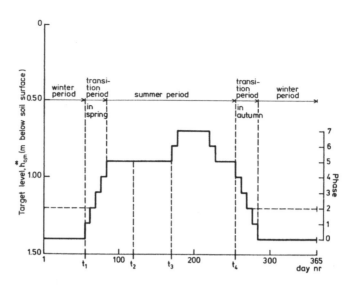

Fig. 5   Schematic representation of the course in time of the target
level and indication of most important periods

timization problem. The lower the surface water level in winter, the better
the drainge but the smaller the possibilities for water conservation. The
higher the surface water level during the growing season the higher the effects of sub-irrigation but the higher the risks of waterlogging. Application of optimization techniques, however, were a priori rejected because
the system at hand is dynamic and highly non-linear. Instead an empirical
approach has been applied. A historical series of weather conditions has
been simulated and by a trial and error procedure the operational rules
have been improved. The result of this procedure can be summarized as follows (see also Figure 5):

- the general trend in the target level is: low in winter and high during
  the growing season;
- the great number of possible target levels is restricted to a limited number by introducing phases with steps of 0.10 m to reduce the frequency of
  adjusting the weir;
- the time to raise the water level in spring (t_1 in Figure 5) depends on
  the groundwater depth. As soon as the (simulated) groundwater depth drops
  below a certain level, the surface water level is raised;
- from t_1 onwards for each higher surface water level to be allowed the
  groundwater depth has to be deeper;
- the frequency of changing the target level is restricted to once per 7 days;

248

- the highest surface water levels are only allowed if the root zone has dried out to a certain (specified) degree and/or is only allowed if $t > t_3$;
- water supply is only possible if $t > t_2$;
- the water supply rate is calculated by the model from the water balance of the surface water system and is limited by the (specified) maximum supply capacity;
- the timing of the beginning of the transition period in late summer/ autumn (t_4 in Figure 5) depends on the dryness of the root zone. The drier the root zone the later the beginning of lowering the surface water. However, t_4 never can surpass a certain date.

The establishment of the different parameters to get a proper way of manipulating the surface water level has been done by simulating, after each change, a historical series of years for the alternatives:

- weir with a fixed crest (no manipulation possible);
- manipulation with the weir but no water supply possible (water conservation);
- manipulating the weir plus water supply possible.

The first alternative has been defined as the reference situation. Next the average yearly effect of water conservation or water supply on crop transpiration and/or the average yearly water supply efficiency has been determined. This trial and error procedure turned out to be successful because the effects of water conservation or water supply in general were not very sensitive for refinements in the operational rules.

3.3. Simulation results

Sensitivity analysis

After having established the operational rules first of all a sensitivity analysis has been carried out. The most important results of this analysis were:

- the influence of soil-physical properties is rather high;
- the drainage resistance has little effect in water conservation and its effect on water supply is somewhat higher;
- the influence of the magnitude of the parameters for reduction in transpiration due to too wet conditions in the root zone is very large. Because the knowledge about these parameters is very poor, it is better to apply a more restrictive type of manipulating the surface water level in the

sense that high surface water levels are only allowed if the groundwater depths are some 0.20 m below those used in the 'normal' type of water level manipulation.

Changes in the hydrological regime, caused by surface water manipulation

The impact of changes in surface water depths on groundwater depths and the water balance of the unsaturated zone depend mainly on drainage resistance and soil-physical properties. To illustrate these impacts one soil-physical/hydrological unit which is representative for the average hydrological conditions in the research area is used.

In Table 1 the water balance of the unsaturated zone for the growing season, averaged over the simulation period 1971-1982 is given. Compared with the reference situation, the lower boundary flux, v_b, increases from -66.6 to -42.3 mm due to water conservation. This extra flux reduces the decrease of the water content in the transition zone, ΔW_s (from -81.5 to -68.3 mm) and does increase actual transpiration, E_t, from 258.2 to 272.2 mm. With water supply, v_b even becomes positive, resulting in a considerably wetter subsoil and an increase in transpiration of 9.2 mm compared with conservation only.

The water balance of the surface water system is presented in Table 2. The most important impact of water conservation and water supply is the decrease in flux density to the ditches, v_d. The effects of manipulating sur-

TABLE 1   Water balance of the unsaturated zone during the growing season (1 April to 1 October) and groundwater depth at 1 October for the water management alternatives: weir with a fixed crest (reference situation), conservation and conservation + water supply; averaged over the period 1971-1982

	Reference situation (no manipulation possible)	Conservation	Conservation + water supply
Net precipitation, P_n (mm)	302.6	302.6	302.6
Soil evaporation, E_s (mm)	57.2	57.2	57.2
Actual transpiration, E_t (mm)	258.2	272.2	281.4
Change in amount of water in root zone, ΔW_r (mm)	2.1	-0.8	-0.1
Change in amount of water in subsoil, ΔW_s (mm)	-81.5	-68.3	-25.7
Lower boundary flux, v_b (mm)	-66.6	-42.3	+10.2
Groundwater depth, h_f^* at 1/10 (m below soil surface)	1.57	1.47	1.25

TABLE 2   Water balance of the surface water system during the growing season (1 April to 1 October) for the management alternatives: weir with a fixed crest (reference situation), conservation and conservation + water supply; averaged over the period 1971-1982

	Reference situation (no manipulation possible)	Conservation	Conservation + water supply
Change in amount of water in tertiary system, $h_{o,t} \cdot a_t$ (mm)	+10.9	+14.4	+8.5
Change in amount of water in secondary system, $h_{o,s} \cdot a_s$ (mm)	+1.2	+1.8	+1.1
Groundwater flux to the tertiary system, v_d (mm)	66.6	42.3	-10.2
Flux over weir, $v_{o,w}$ (mm)	-78.7	-58.5	-72.6
Flux from inlet structure, $v_{o,p}$ (mm)	-	-	73.2

face water levels on groundwater depths are illustrated in Figure 6, in which for three selected years the simulated depths for the reference situation, water conservation and conservation + water supply are depicted. This figure clearly illustrates that water conservation retards the drop in groundwater depth during the growing season, especially in the dry years 1971 and 1982. Also in this type of years the effect of water supply on groundwater depth in the growing season is considerable. On the contrary, in wet years like 1980, the effect of water conservation is limited while the effect of water supply is negligible.

Effects of manipulating the surface water level for the entire area 'De Monden'

To arrive at the average yearly effects of surface water manipulation on transpiration for the entire area 'De Monden' the following steps have been taken:

- the area has been divided into 20 subregions, i.e. the areas which constitute more or less the drainage basin of the different adjustable weirs (see also Figure 2);
- For each subregion one relationship $v_a = f(h_f^*, h_o^*)$ has been calculated with the FEMSATS-model. This relationship is the average of the $v_a = f(h_f^*, h_o^*)$-relationship of each nodal point of the particular subregion;
- for each subregion a representative drainage resistance and soil-physical

251

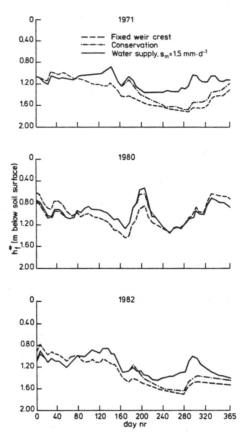

Fig. 6 Effects of different ways of surface water management on the groundwater depths during 1971 (moderate dry growing season), 1980 (very wet period around end of July) and 1982 (rather dry year)

unit has been selected. For the representative soil-physical/hydrological units the transpiration of a potato crop during the period 1971-1982 has been simulated with the SWAMP-model for the alternatives:

I   weir with a fixed crest = reference situation

II   water conservation

III water conservation plus water supply with a maximum supply flux density, s_m, of 0.75 mm·d^{-1}

IV   as III, s_m = 1.50 mm·d^{-1}

V   as III, s_m = 2.50 mm·d^{-1}

- from these simulations the gross effects per unit area of water conservation (II-I) and water supply (III-II, etc.) have been derived. Note that for calculating the effects of water supply the reference situation is water conservation;

- the gross effects per unit area have been corrected for the limited number of hydrological years, the crop rotation scheme, the gross-net area, the undulation of the soil surface and the expected changes in soil-physical properties, to arrive at the net hydrological effects per unit area per subregion;
- the net hydrological effects can be distinguished in effects on average yearly transpiration, $\overline{\Delta T}$, and (in case of water supply) average water supply efficiency, $\overline{e}$. The latter has been defined as the quotient of average yearly change in transpiration and average yearly amount of supplied water per unit area, $\overline{a}$;
- table 3 gives the results.

Because of the relatively wet climate in the Netherlands the average yearly effects of water conservation and water supply are limited. However, in dry years the effect of water conservation is about two times higher than in normal years while the effect of water supply is about threefold. In these dry years there remains a reduction in transpiration, even with water supply for sub-irrigation. Therefore water supply by sub-irrigation can be characterized as an additional way of irrigation.

Although the average efficiency of water supply by sub-irrigation is low, again, in dry years, this efficiency is about threefold.

4. ECONOMICAL ANALYSIS OF EFFECTS OF SURFACE WATER MANAGEMENT

4.1. Benefits of surface water management

The net hydrological effects of water conservation and water supply, as given in Table 3, have been transformed in financial effects to obtain the increase in income of the farmers in the area. For this purpose the net hy-

TABLE 3   Net hydrological effects per unit area of water conservation and conservation + water supply with different maximal supply flux densities, averaged over the entire region 'De Monden'

	Water conservation	Conservation + water supply		
		s_m = 0.75	s_m = 1.50	s_m = 2.50 mm·d^{-1}
$\overline{\Delta T}$ (mm·y^{-1})	6.0	3.7	5.4	6.3
ΔT (%)	2.02	1.25	1.82	2.14
$\overline{a}$ (mm·y^{-1})	–	39.5	59.4	67.2
$\overline{e}$ (%)	–	9.4	9.1	9.4

TABLE 4 Average yearly increase of agricultural income (10^3 Dfl) due to water conservation and conservation + water supply with different max. supply flux densities

Water conservation	Conservation + water supply		
	$s_m = 0.75$	$s_m = 1.50$	$s_m = 2.50$ mm·d^{-1}
520	318	464	547

drological effects have been transformed into effects on crop production. On the basis of actual yields and prices of products it has been calculated that 1% increase in actual transpiration means Dfl 48.25 per ha arable land. On this basis the average yearly increase in agricultural income for the entire area has been computed. The results are presented in Table 4.

4.2. Costs of surface water management

To induce the increase in agricultural income, the waterboard and the provincial and national authorities have to invest and have to make costs for operation and maintenance of the surface water system.

To have adequate drainage, farmers will take care of an effective tertiary system. This system can be used for conservation and water supply without any additional investments or variable costs.

To ensure water supply to the different districts in the Netherlands, the national water authority certainly has to invest or will make variable costs. In the framework of the PAWN-study (Policy Analysis of the Water

TABLE 5 Total investments and average yearly variable costs (10^3 Dfl) of water conservation and conservation + water supply to be made by the provincial water authority and the waterboard 'De Veenmarken', in prices of 1980

	Water conservation	Conservation + water supply		
		$s_m = 0.75$	$s_m = 1.50$	$s_m = 2.50$ mm·d^{-1}
Provincial water authority				
Investments		1280	2540	4230
Variable costs		77	116	157
Waterboard				
Investments	600	691	691	691
Variable costs	9	11	12	12

Management in the Netherlands) a model for the optimal water distribution has been developed (Abrahamse et al., 1982). At this moment, however, water from the national water system is free of charges, so in this study no costs of surface water management on national level have been taken into account.

The data on total investments for water conservation and for water supply and the variable costs for the drainage-district are presented in Table 5.

4.3. Internal rate of return of surface water management projects

The internal rate of return, i (-), is such that

$$\sum_{n=1}^{N} \frac{B(n)}{(1+i)^n} - \sum_{n=1}^{N} \frac{I(n) + C_v(n)}{(1+i)^n} - \frac{R}{(1+i)^N} = 0 \tag{7}$$

where N is lifetime of the project under consideration (years), B(n) is agricultural benefits of the project in year n (Dfl), I(n) is investments in the project in year n (Dfl), C_v is variable costs of the project in year n (Dfl) and R is rest value of the surface water management facilities in year N. For lifetime N is taken 30 years with R is zero. Because of from year to year varying weather conditions B(n) and also $C_v(n)$ will vary considerably. It has been assumed that it is allowed to take B(n) and $C_v(n)$ to be equal to the average yearly benefits and costs as given in Tables 4 and 5. Now the internal rate of return of the projects for water conservation and water supply can be calculated. Table 6 shows the results.

TABLE 6   Internal rate of return (-) of water conservation and water supply projects for the region 'De Monden'

Water conservation	Water conservation + water supply		
	s_m = 0.75	s_m = 1.50	s_m = 2.50 mm·d^{-1}
0.49	0.12	0.10	0.08

Discussion

The internal rate of return of the project for water conservation is very high. The figure of 0.49 implies that the investments in water conservation pay back in two years.

The internal rates of return of water supply projects are quite acceptable. In the Netherlands an internal rate of return of more than 0.10 is

considered to be high. The figure of 0.08 for investments in water supply
with a maximum supply flux density to the region of 2.50 $mm \cdot d^{-1}$ does not im-
ply that the internal rate of return of e.g. an increase in max. supply flux
density from 2.2 to 2.5 $mm \cdot d^{-1}$ is also 0.08.

4.4. <u>Maximum supply flux density as function of internal rate of return</u>

From the figures in Tables 4 and 5 relationships have been derived between benefits, variable costs and investments respectively and maximum supply flux density. The derivative of these functions gives the marginal benefits, variable costs and investments as a function of the supply capacity.
Next with the use of eq. (7) the internal rate of return can be calculated
as a function of the supply capacity.

Investments in water supply projects usually are done in finite steps,
whereas the first step in general is chosen in such a way that the internal
rate of return begins to decrease with an increase in supply capacity. Of
each step the benefits, variable costs and investments can be derived using
the data of Tables 4 and 5 and consequently the internal rate of return of
the increase in supply capacity can be calculated. The results of this procedure are presented in Figure 7. From this figure the supply capacity to be
installed with a particular choice of the internal rate of return can be determined directly. On the other hand for any proposed supply capacity the
figure provides the marginal internal rate of return.

5. APPLICABILITY TO OTHER REGIONS

Whether or not projects for water conservation and water supply for sub-
irrigation are necessary depends to a great extent on the hydrological and
topographical properties of the region under consideration. The most impor-

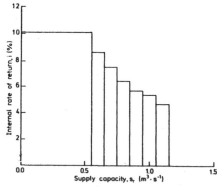

Fig. 7   Internal rate of return of increase in supply capacity to the
region 'De Monden' of 0.1 $m^3 \cdot s^{-1}$, as function of total maximum supply
rate

256

tant properties in this respect are the interaction between surface water and groundwater system (i.e. the drainage resistance), the capillary properties of the soil, the intensity of seepage (i.e. the groundwater depth) and the degree of undulation of the soil surface. Furthermore the height of necessary investments and variable costs are of utmost importance. Under Dutch circumstances, the research area can be classified as favourable with respect to drainage resistance and capillary conductivities, but unfavourable with respect to height of investments in water supply, because the pumping height is appr. 16 m.

REFERENCES

Abrahamse, A.H., Baarse, G. and Van Beek, E. 1982. Policy analysis of water management in the Netherlands. Vol. XII. Model for regional hydrology, agricultural water demands and damages from drought and salinity. 315 pp.

Belmans, C., Wesseling, J.G. and Féddes, R.A. 1983. Simulation model of the water balance of a cropped soil: SWATRE. J. of Hydrol. 63(3/4): 271-286.

Bloemen, G.W. 1980a. Calculation of hydraulic conductivities of soils from texture and organic matter content. Zeitschr. für Pflanzenernähr. Bodenk. 143,5: 581-605.

Bloemen, G.W. 1980b. Calculation of steady state capillary rise from the groundwater table in multi-layered soil profiles. Zeitschr. für Pflanzenernähr. Bodenk. 143,6: 701-719.

Boels, D., Van Gils, J.B.H.M., Veerman, G.J. and Wit, K.E. 1978. Theory and system of automatic determination of soil moisture characteristics and unsaturated hydraulic conductivities. Soil Sci. 126,4: 191-199.

Ernst, L.F. 1962. Groundwater flow in the saturated zone and its calculation with the presence of horizontal open watercourses. PUDOC, Wageningen. 189 pp. (in Dutch)

Laat, P.J.M. De. 1980. Model for unsaturated zone above a shallow water table, applied to a regional sub-surface flow problem. PUDOC, Wageningen. 126 pp.

Nieuwenhuis, G.J.A., Smidt, E.H. and Thunnissen, H.A.M. 1985. Estimation of regional evapotranspiration of arable crops from thermal infrared images. Int. J. of Remote Sensing (in press).

Querner, E.P. 1984. Program FEMSAT. Part 1. Calculation method for steady and unsteady groundwater flow. Nota 1557 ICW, Wageningen. 24 pp.

Smidt, E.H. 1984. Application of the stationary groundwater flow model FEMSATS in 'De Monden'. Nota 1515 ICW, Wageningen. 53 pp. (in Dutch)

The flow rate of water at the level of a moving water table

A.POULOVASSILIS & S.AGGELIDES
Agricultural University, Athens, Greece

ABSTRACT

A semi-empirical analysis is developed for the calculation of the flow rate across a moving water table and of the specific yield (or of the specific uptake). The flow rate and the specific yield (or uptake) were determined experimentally in a vertical sand column for various constant speeds of ascent or descent of the water table, for various initial positions of the water table with respect to the upper surface of the column and for various soil water profiles prevailing in the column before the initiation of the water table movement. It was found that the flow rate depends on the time elapsed since the initiation of the water table movement as well as on the magnitude of the speed at which the water table is moving, on the initial position of the water table and on the soil water profile prevailing initially in the column. Moreover it was found that the specific yield (or uptake) depends on the time elapsed since the initiation of the water table movement as well as on the initial position of the water table and on the soil water profile prevailing in the porous column before the initiation of the water table movement.

1. INTRODUCTION

It has been shown (Childs and Poulovassilis, 1962) that the shape of the soil water profile developing above a moving water table differs from that of the steady-state water content profile. In particular, it has been shown that the former profile when compared with the latter, appears to be compressed when developed above an ascending water table and extended when developed above a descending water table and that the differences in shape depend on the velocity of the moving water table. Furthermore that when the water table ascends or descends with a constant speed, the water content profile acquires a constant shape which can be determined.

In the above work it was recognized that the transition from the steady-state water content profile, prevailing before the initiation of the water table movement, to the moving water content profile of constant shape, includes a series of intermediate water content profiles whose development is not possible to follow analytically. Thus, the flow rate at the water table cannot be calculated. The present work aims in presenting some experimental results which show the dependency of the flow rate at the water table on the speed with which the water table ascends or descends and on the nature of the initial steady-state water content profile. Moreover, a semi-empirical analysis for the calculation of the flow rate at a water table moving with a constant speed is presented.

2. THEORY

Let us assume that the water content profiles of Figure 1 are constant shape profiles that have been developed above a water table and maintained by a constant flow rate q, positive downwards, applied at the soil surface. In particular, let us assume that profiles (a), (b) and (c) represent a steady-state water content profile, a descending water content profile and an ascending water content profile respectively.

When the relationships between the soil water pressure head P and the soil hydraulic conductivity K and between P and the soil water content θ are known, then the above profiles can be determined (Childs, 1956; Youngs, 1957). Then:

$$z = \int_0^P \frac{dP}{q/K-1} \tag{1}$$

gives the steady-state pressure profile (Childs and Poulovassilis, 1962) and

$$z = \int_0^P \frac{dP}{q/K-1 - (V/K)(\theta-\theta_u)} \tag{2}$$

gives the moving pressure head profile subject to the condition

$$V > -(dK/d\theta)_{\theta_u} \tag{3}$$

where V is the constant speed of descent or ascent of the water table taken positive upwards and θ_u the soil water content at K = q. Condition (3) is satisfied for all rising water tables and for all water tables which fall at a speed smaller than the magnitude of the term at the right hand side of (3).

The corresponding water content profiles to those given by eqs. (1) and (2) are obtained by using the moisture characteristic of the soil. Therefore the amount of water can be calculated from:

$$S_\infty = \int_0^\infty (\theta_m-\theta)dz \tag{4}$$

where θ pertains to the steady-state water content profiles θ_m to the moving water content profile, and S_∞ is the difference in stored water above the moving water table per unit cross-section due to the changes of the shape of the water content profile occurring after a sufficiently long period of movement of the water table.

The amount of water Q which has crossed the water table from the initiation of the water table desplacement (t = 0) up to time t, is given by:

$$Q = (\theta_u - \theta_o)Vt - S + qt \tag{5}$$

where θ_o is the water content of the column at saturation and S the amount of water which has been stored above the water table or the amount of water which has been released as a consequence of the changes of the water content profile up to time t. For the determination of the value S it can be assumed that the rate of change of S is proportional to the difference between its final value S_∞ when $t \to \infty$ and the value S at time t, i.e.

$$\frac{dS}{dt} = a(S_\infty - S) \tag{6}$$

where a is a proportionality constant. The integration of eq. (6) gives:

$$-\ln(S_\infty - S) = at + C \tag{7}$$

From the initial condition t = 0, S = 0 it follows that $C = -\ln S_\infty$ so that eq. (7) becomes:

$$S = S_\infty(1 - e^{-at}) \tag{8}$$

Substituting this value of S in eq. (5) and subsequent differentiation with respect to time gives:

$$\frac{dQ}{dt} = (\theta_u - \theta_o)V - aS_\infty e^{-at} + q \tag{9}$$

where dQ/dt represents the flow rate at the moving water table at time t. If

$$(dQ/dt-q) << (\theta_u - \theta_o)V$$

then from (9):

$$(\theta_u - \theta_o)V = aS_\infty e^{-at} \tag{10}$$

which for t = 0 gives:

$$a = V/S_\infty(\theta_u - \theta_o) \tag{11}$$

Thus eq. (8) becomes:

261

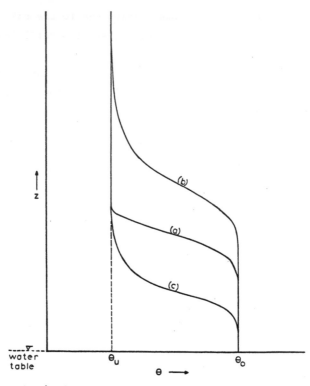

Fig. 1   Hypothetical constant shape water content profile: (a) steady-state water content profile, (b) descending water content profile, (c) ascending water content profile

$$S = S_\infty(1 - e^{-V/S_\infty(\theta_u-\theta_o)t}) \tag{12}$$

and eq. (5):

$$Q = (\theta_u-\theta_o)Vt - S_\infty(1 - e^{-V/S_\infty(\theta_u-\theta_o)t}) + qt \tag{13}$$

Eq. (13) gives the total amount of water which has crossed the moving water table to time t. When $t \to \infty$, eq. (9) becomes:

$$\frac{dQ}{dt} = (\theta_u - \theta_o)V + q \tag{14}$$

Let us assume next, that after the development of a descending water content profile, the water table displacement is ceased and remains stationary thereafter. As a result the constant shape water content profile will gradually become a steady-state profile and a certain amount of water, S', will finally be released from the column (see Figure 1). On the other hand in the case of an ascending water table a steady-state water content profile

262

will develop because of an increase of the water stored above the water table by an amount S'. The amount of water Q' which has crossed the stationary water table from the cessation of its movement (t' = 0) up to the time t' is given by the equation:

$$Q' = S' + qt' \tag{15}$$

where S' is the amount of water which has been released or absorbed by the porous column due to the changes in shape of the water content profile. Again one may assume that:

$$\frac{dS'}{dt'} = a'(S'_\infty - S') \tag{16}$$

from which

$$S' = S'_\infty(1 - e^{-a't'}) \tag{17}$$

Substitution of S' into eq. (15) and differentiation with respect to time give:

$$\frac{dQ'}{dt'} = S'_\infty a' e^{-a't'} + q \tag{18}$$

Because the flow rate at the water table cannot be discontinued, it follows that for $t \to \infty$ and $t' \to 0$:

$$\frac{dQ'}{dt'} \to \frac{dQ}{dt} \tag{19}$$

and therefore eqs. (14) and (18) give:

$$a' = V/S'_\infty(\theta_u - \theta_o) \tag{20}$$

Thus eq. (17) becomes:

$$S' = S'_\infty(1 - e^{-V/S'_\infty(\theta_u - \theta_o)t'}) \tag{21}$$

and eq. (15):

$$Q' = S'_\infty(1 - e^{-V/S'_\infty(\theta_u - \theta_o)t'}) + qt' \tag{22}$$

The last equation allows the calculation of the amount of water which has crossed the stationary water table during the time interval t' = 0 to t'.

3. EXPERIMENTAL PROCEDURE AND RESULTS

For the development of the water content profiles and the measurements

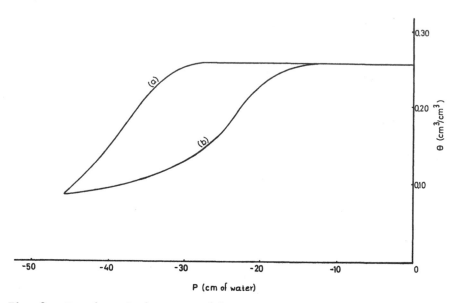

Fig. 2   Boundary drying curve (a) and boundary wetting curve (b) of the relationship between soil water pressure head P and water content θ for the experimental column

of Q and Q', a homogeneous column of length 2.10 m was used. For the construction of the column, sand fractions were packed in a transparent perspex tube of internal diameter 28 mm and wall thickness 6 mm, in such a way as to achieve homogeneity of the column throughout its length. Figures 2 and 3 show the P-θ and K-θ relationships for the porous materials used. These relationships were obtained by the method described elsewhere (Poulovassilis, 1970).

Tensiometers were installed along the whole length of the column at distances of 5 and 10 cm and connected via a hydraulic switch to a pressure transducer for automatic recording. In this way it was possible to follow the pressure head profiles continuously. By translating them into water content profiles through the P-θ relationship, these profiles could be followed too.

Two channels were cut in the wall of the perspex tube along its whole length and covered with a wire gauge so that free contact of the porous body with the atmosphere as well as the control of the water table were ensured. The channels were connected via a flexible U tube (see Figure 4) to a small reservoir Δ, in which the level of water was maintained constant.

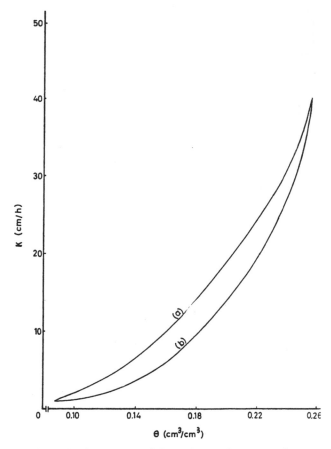

Fig. 3   Boundary drying curve (a) and boundary wetting curve (b) of the relationship between water content θ and hydraulic conductivity of the column

The reservoir Δ was moved along a vertical screw rod which was rotated by an electric motor and a variable gear-box. In this way the reservoir could move upwards and downwards with different constant speeds. When the water table level was falling (descent of the reservoir Δ) the draining water was collected in a series of burettes. During the rise of the water table the water level in the reservoir Δ was kept constant through a graduated Mariotte tube which was moving along with the reservoir Δ. By this means the water quantity taken up by the column could be measured.

In Figure 5 curves of cumulative outflow versus time are presented for various constant speeds of water table descent. Before the initiation of the water table descent the water table was, for all cases, at a depth of 70 cm below the surface of the column. This depth was sufficient as to allow the

265

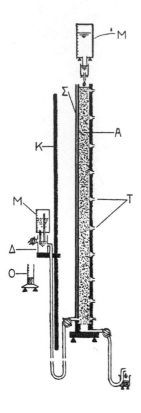

Fig. 4   The experimental arrangement: (Δ)
reservoir, (K) vertical rod with screwing
threads, (M) Mariotte tube, (0) volumetric
tube, (T) tensiometers, (Σ) channels for air
flow and water table control

development, at its full length, of a steady-state water content profile
sustained by a constant rainfall q = 0.7 cm/h. This water content profile
together with the profiles of constant shape developed for various constant
speeds of water table descent are shown in Figure 6. It should be noted
that the applied speeds were smaller than the speed allowed by condition
(3) which for θ_u = 0.086 was 65 cm/h (see Figure 3).

In Figure 7 curves of cumulative outflow versus time for the same speed
of water table descent (V = -19.6 cm/h) are presented. In these curves the
effect of the distance of the initial water table from the soil surface is
shown. For curve (a) the distance 28.5 cm only allowed the development of
the wetter part of the drying water content profile, while for curve (b)
the distance 70 cm allowed the full development of the water content profile. In both cases the rainfall rate was q = 0.7 cm/h.

Figure 8 presents cumulative outflow curves obtained for the same speed
of water table descent (V = -26.8 cm/h) starting from the same water table
position (70 cm below the soil surface). These curves refer to different
initial water content profiles. So curve (a) refers to an initially drying

266

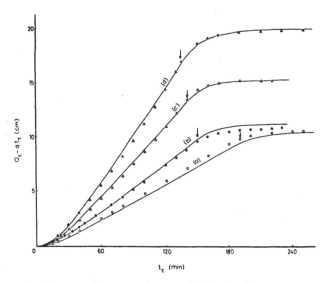

Fig. 5 Cumulative outflow curves. Initial drying water content profiles for q = 0.7 cm/h. Initial distance of water table from soil surface 70 cm. Constant speed of water table descent, (a): V = -19.6 cm/h, (b): V = -26.8 cm/h, (c): V = -39.4 cm/h, (d): V = -53.5 cm/h. Full lines are calculated curves using eqs. (13) and (22). Triangles and circles (open and shaded) are experimental points. The vertical arrows show the time of cessation of water table movement. Q_r and t_r denote the cumulative outflow and the time during the water table movement and after its cessation

water content profile, while curve (b) refers to an initially wetting water content profile. These initial water content profiles are shown in Figure 9.

Figure 10 shows cumulative inflow curves, obtained during rising from the same position with V = 19.6 cm/h. Curve (a) corresponds to an initially drying water content profile while curve (b) corresponds to an initially wetting water content profile.

In Figures 11, 12 and 13 calculated curves of specific yield (eq. 24) are compared with experimental values obtained from the cumulative outflow curves shown in Figures 5, 7 and 8, and in Figure 14 calculated curves of specific uptake (eq. 24) are compared with experimental values obtained from the curves of cumulative inflow shown in Figure 10.

4. DISCUSSION

Comparison of experimental and calculated curves in Figures 5, 7, 8 and 10 shows that the semi-empirical analysis developed in the present work allows satisfactory calculation of the flow rate across a moving water table

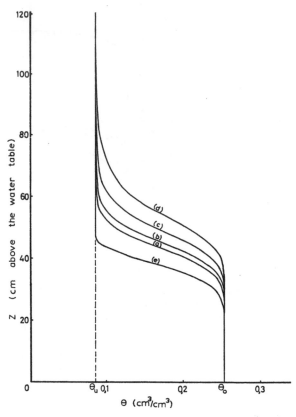

Fig. 6   Drying water content profiles of constant shape developed above
a descending water table sustained by a constant rate of precipitation
q = 0.7 cm/h, (a): V = -19.6 cm/h, (b): V = -26.8 cm/h, (c): V = -39.4
cm/h, (d): V = -53.5 cm/h, (e): V = 0. Moisture profile (e) represents
the initial drying profile before the initiation of the water table
movement

as well as that after the cessation of the movement.

The differences between the curves shown in Figure 5 illustrate the dependency of the cumulative outflow Q on the speed of descent. In particular,
from Figure 5 it may be seen that increase of the speed of fall causes a
general increase in the slopes of the curves, i.e. an increase of the flow
rate across the moving water table and therefore an increase of Q according
to eqs. (9) and (13) for the time interval of descent, and according to eqs.
(18) and (22) for the time interval following the cessation of the water table movement.

The flow rate at the level occupied momentarily by the water table at
time t may be considered as the sum of the constant flow rate q applied at
the soil surface and of the flow component resulting from the release or

268

uptake of water by the unsaturated zone above the moving water table. Thus, if we denote by μ the volume of water, released or absorbed by the unsaturated zone per unit cross-sectional area per unit change of the water table level, then

$$\frac{dQ}{dt} = \mu \frac{dz}{dt} + q \tag{23}$$

Comparison of eq. (23) and eq. (9) gives:

$$\mu = (\theta_u - \theta_o)(1 - e^{-V/S_\infty(\theta_u - \theta_o)t}) \tag{24}$$

The quantity μ is usually known as specific yield for a falling water table or as specific uptake for a rising water table. From eq. (25) it is obvious that when $t \to 0$, $\mu \to 0$ and when $t \to \infty$, $\mu \to (\theta_u - \theta_o)$. The value of μ for given $(\theta_u - \theta_o)$ and t values depends on the magnitude of the ratio (V/S_∞) in the sense that an increase or decrease of this value causes an increase or decrease of μ. Figure 11 shows μ as a function of time for two different values of the ratio (V/S_∞). Figure 5 showed that the quantity of water which is finally released from the column after the descending water table is brought to a halt is an increasing function of V. As already mentioned (see eq. 2), the fall of the water table causes an extension of the water content profile, which becomes more pronounced as V increases (see Figure 6). As a result, the quantity of water retained into the unsaturated zone increases as V assumes greater values. This quantity, which is called delayed yield is gradually released after the cessation of the water table fall.

The effect of the position of the water table relative to the soil surface on the cumulative outflow and on the specific yield, has been already pointed out by Childs (1960). This effect is shown clearly in Figures 7 and 12. In the case of curve (a) the initial position of the water table was 28.5 cm below the soil surface so that only part of the steady-state water content profile was allowed to develop, whereas in the case of curve (b) the initial position of the water table was 70 cm below the soil surface and the steady-state water content profile was fully developed. The difference in cumulative outflow and μ shown by Figures 7 and 12 is quite considerable and could have been bigger if the water table had been closer to the soil surface. The analysis, presented in this study takes into consideration the effect of the initial water table position by giving to S_∞ its proper value. Thus, for curve (a) $S_\infty = 2.548$ cm, whereas for curve (b) $S_\infty =$

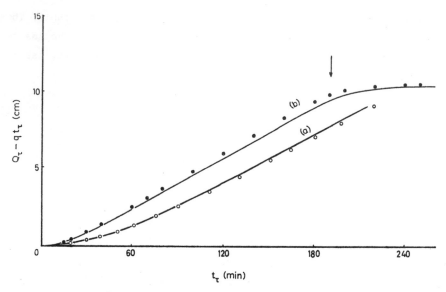

Fig. 7 Cumulative outflow curves. Initial drying water content profiles for q = 0.7 cm/h. Speed of descent of the water table V = -19.6
cm/h. Initial distance of the water table from the column surface (a)
28.5 cm and (b) 70 cm. Full lines are calculated curves using eqs. (13)
and (22). Open and shaded circles are experimental points

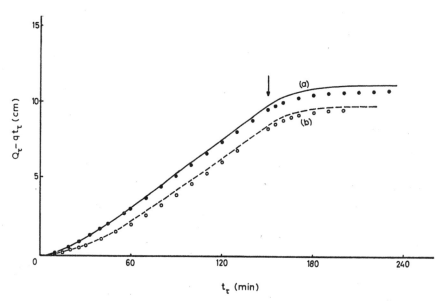

Fig. 8 Cumulative outflow curves. Speed of descent of the water table
V = -26.8 cm/h. Initial distance of the water table from the surface 70
cm, q = 0.7 cm/h. (a) initial drying water content profile, (b) initial
wetting water content profile. Full lines are calculated curves using
eqs. (13) and (22). Open and shaded circles are experimental points

270

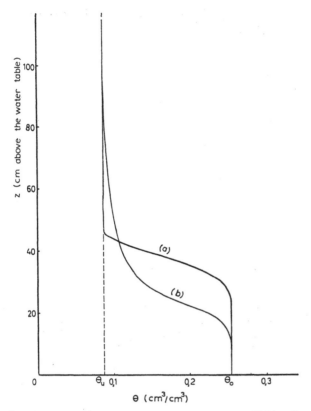

Fig. 9   Steady-state water content profiles developed under constant rate of precipitation q = 0.7 cm/h. (a) drying water content profile, (b) wetting water content profile

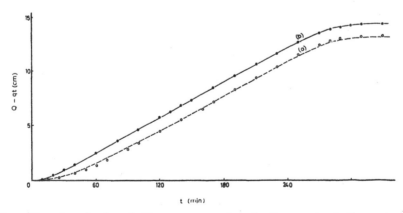

Fig. 10   Cumulative inflow curves. Speed of ascent of the water table V = 19.6 cm/h. Initial distance of water table from the surface 117 cm. Initial steady-state water content profile with q = 0.7 cm/h. (a) drying, (b) wetting. Full lines are calculated curves using eq. (13) and open and shaded circles are experimental points

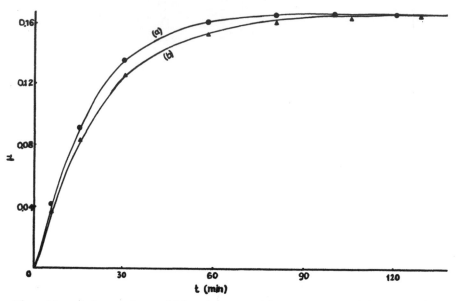

Fig. 11   Curves of specific yield μ. (a): V = -26.8 cm/h, S_∞ = 1.378 cm, V/S_∞ = -16.517 h^{-1}, (b): V = -53.5 cm(h, S_∞ = 3.239 cm, V/S_∞ = -16.517^{-1}. Full lines are calculated curves using eq. (24). Shaded circles and triangles are values determined from the experimental cumulative outflow curves (b) and (d) of Figure 5

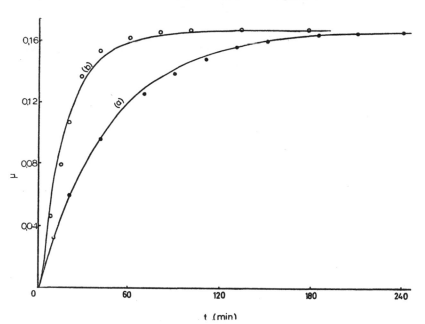

Fig. 12   Curves of specific yield μ. (a) distance from the surface 28.5 cm, S_∞ = 2.548 cm, V/S_∞ = -7.692 h^{-1}, (b) distance from the surface 70 cm, S_∞ = 1.020 cm, V/S_∞ = -19.215 h^{-1}. Full lines are calculated curves using eq. (24). Open and shaded circles are values determined from the experimental curves (a) and (b) of Figure 7

272

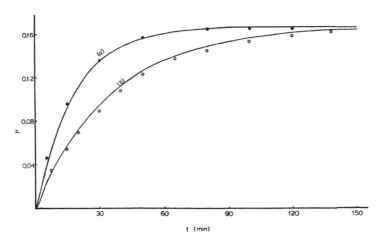

Fig. 13   Curves of specific yield μ. (a) initial drying water content profile. $S_\infty = 1.378$ cm, $V/S_\infty = -19.448$ h^{-1}, (b) initial wetting water content profile, $S_\infty = 2.715$ cm, $V/S_\infty = 9.871$ h^{-1}. Full lines are curves calculated using eq. (24). Open and shaded circles are values determined from the experimental curves (a) and (b) of Figure 8

1.020 cm. It is evident that the increase of S_∞ causes a decrease of the cumulative outflow and of μ according to eqs. (13) and (24).

The effect of hysteresis on the cumulative outflow and on specific yield for a descending water table is shown in Figures 8 and 13. On account of the descending water table, the wetting water content profile (see Figure 9), prevailing initially in the porous column, is gradually transformed into a descending drying water content profile. As a result a greater amount of water is stored above a falling water table than in case the initially steady-state profile was a drying one. For this reason, the cumulative outflow shown by curve (b) of Figure 8 lacks considerably that shown by curve (a). Similarly, the specific yield shown by curve (b) of Figure 13 is considerably smaller than that of curve (a) for small values of time. The value of S_∞ for curve (a) was 1.378 cm and for curve (b) 2.715 cm. Substitutions of these values into eqs. (13) and (24) led to satisfactory calculation of Q and μ (see Figures 8 and 13).

The cumulative inflow curves shown in Figure 10 present a gradual decrease of slopes for larger times. This is due to the fact that at those times the water table was approaching the soil surface. The differences between the curves (a) and (b) in Figures 10 and 14 are again due to hysteresis in the P-θ relationship.

273

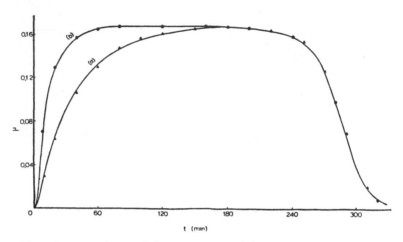

Fig. 14   Curves of specific uptake μ. (a) initial drying water content profile, $S_\infty = -2.105$ cm, $V/S_\infty = -9.311$ h^{-1}, (b) initial wetting water content profile, $S_\infty = -0.768$ cm, $V/S_\infty = -25.520$ h^{-1}. Full lines are curves calculated using eq. (24). Shaded circles and triangles are values obtained from the experimental curves (a) and (b) of Figure 10

REFERENCES

Childs, E.C. 1956. Recent advances in the study of water movement in unsaturated soil. Sixième congrès de la science du sol, Paris.

Childs, E.C. and Poulovassilis, A. 1962. The moisture profile above a moving water table. J. Soil Sci. 13: 272-285.

Poulovassilis, A. 1970. Hysteresis of pore water in granular porous bodies. Soil Sci. 109: 5-12.

Youngs, E.C. 1957. Moisture profile during vertical infiltration. Soil Sci. 84: 283-290.

Discussion Session 4

Question from Mr. G.A. Oosterbaan to Mr. P. Bazzoffi

    To what period of time do the siltation data which have been used for
the assessment of the statistical relationships you discussed, refer to?

Answer of Mr. Bazzoffi

    We used different periods for the different reservoirs. For instance,
we surveyed siltation in three reservoirs in Tuscany over a period of
19-25 years, while for the other reservoirs we have data over periods
ranging from a few years to more than 40 years.

Question from Mr. G.A. Ven to Mr. S. Aggelides

    Why have experiments been used instead of models for describing the flow
situation?

Answer of Mr. Aggelides

    A semi-empirical analysis was followed to obtain an insight of what
physically happens with the flow above a moving water table. Research
of this kind must be continued to acquire more knowledge on the soil-
water phenomena and more detailed information about the necessary input
in models.

Question from Mr. A. Armstrong to Mr. P.J.T. van Bakel

    I would like to ask about the need for level ground surfaces in sub-
irrigation systems. Our experience with the use of drainage systems for
sub-irrigation has shown that it is technically feasible to establish a
water table in the soil at whatever level chosen. However, the problem
arises in deciding the appropriate water table level in relation to the
microtopography within each field, which in the UK context can often be
as much as 40-50 cm even in generally flat area. As a result, the first
step in the successful use of this technique is the levelling of the
land surface, which is a sufficiently difficult and costly operation
that it has generally prevented its widespread adoption. Do you have
similar problems in the area you describe?

Answer of Mr. Van Bakel

    Also in the Netherlands levelling is too costly. In the case study, we
had to do with a relatively flat surface so that sub-irrigation was
possible. Due to unevenness of the soil surface a chosen water table
may be too high for low places and too low for high places. The influ-

ence of the unevenness of the soil surface on the final results is rather large so levelling is a prerequisite for sub-irrigation.

Questions from Mr. J. Feyen to Mr. P.J.T. van Bakel

1) How is the reduction in transpiration under too wet conditions taken into account in your model?

2) From the sensitivity analysis you found a rather large influence of soil physical properties. Do you have any idea about the minimum sampling density that is necessary to describe the area properly?

3) What is the order of magnitude of the computer time required to run the model as compared to the computing time needed to run the SWATRE program for one season?

Answer of Mr. Van Bakel

1) The reduction in transpiration is defined as a function of the relative water content in the root zone. The two parameters used in this function are empirical.

2) Soil-physical properties have been derived from soil mapping units, so the soil map is used as the source for soil physical properties. Since we did not use data from samples, I cannot answer this question.

3) The computer time for running one season is about 30 cpu seconds on a VAX 11/750 computer. SWATRE requires about 60 seconds.

Session 5
Effects of agriculture on its environment

Chairman: P.E.RIJTEMA
Institute for Land & Water Management Research (ICW),
Wageningen, Netherlands

Leaching of ferrous iron after drainage of pyrite-rich soils and means of preventing pollution of streams

L.B.CHRISTENSEN & S.E.OLESEN
Hedeselskabet, Danish Land Development Service, Denmark

ABSTRACT

Many Danish soils contain the iron-rich mineral pyrite. When the ground-water level is lowered pyrite will be oxidized and ferrous iron is leached out into the streams. This may have serious impacts on stream ecology.

In 1981 the Danish government passed a bill according to which it was decided to investigate the impact of ochre on stream ecology. Since then the Danish Land Development Service has carried out a number of experiments on how to remove ferrous iron from drainage water. During the period 1981 to 1984 a number of 20 ochre abatement plants were built, financed by the Ministry of the Environment. Both chemical and biochemical methods and techniques of ferrous iron oxidation and sedimentation were investigated.

The investigations showed that treatment of the ferrous drainage water with quick lime is the most efficient and reliable method. When the pH of the water was higher than 7.3 the average ferrous iron removal was 97%. Biochemical methods also showed a relatively high iron removal efficiency (about 70%) but they are not fully developed and the efficiency of these methods can probably be increased to the same high level as the quick lime method.

1. INTRODUCTION

During the last century a large part of the Danish soils have been drained in order to improve them for the use as farmland. In some instances this has led to pollution of streams with ochre. Many Danish soils, especial-ly in the western part of Jutland, contain the iron-rich mineral pyrite in the subsoil. The pyrite containing soils are mainly found in littoral depos-its, late glacial deposits, marsh areas, bogs and meadows, where environ-mental conditions during the deposition and sedimentation favoured the for-mation of pyrite. The substances that are needed in the formation of pyrite are sulphur, iron and organic matter together with anaerobic conditions.

2. OXYDATION OF PYRITE

Under reduced conditions, that is when the pyrite containing soil is saturated with water, pyrite is a non-soluble and stable mineral. Then it accumulates and can be found in concentrations up to 20% in some Danish soils. If, however, the gorundwater level is lowered, pyrite is oxidized. The process is rather complicated, and involves a number of intermediate reactions.

The process can proceed as a pure chemical reaction between oxygen and pyrite. This is, however, a very slow process:

$$FeS_2 + 7/2 \ O_2 + H_2O \rightarrow Fe^{2+} + 2 \ SO_4^{2-} + 2 \ H^+ \qquad (1)$$

The ferrous iron is further oxidized according to:

$$Fe^{2+} + 1/4 \ O_2 + H^+ \rightarrow Fe^{3+} + 1/2 \ H_2O \qquad (2)$$

and the resulting ferric iron precipitates as ochre:

$$Fe^{3+} + 3 \ H_2O \rightarrow Fe(OH)_3 \ (ochre) + 3 \ H^+ \qquad (3)$$

The process can also be a very quick reaction between pyrite and ferric iron:

$$FeS_2 + 14 \ Fe^{3+} + 8 \ H_2O \rightarrow 15 \ Fe^{2+} + 2 \ SO_4^{2-} + 16 \ H^+ \qquad (4)$$

The solubility of ferric iron depends on the pH. At pH levels above 4, the ferric iron will precipitate. Generally the slow process of reaction (2) dominates, and therefore the chemical oxydation of pyrite will normally proceed slowly.

In addition pyrite can be oxidized at a very high rate by bacteria, Thiobacillus. Both the quick chemical reaction (4) and the biochemical oxyda- tion proceed best at low pH levels (between 2 and 4).

Chemical oxydation of ferrous iron by eq. (3) is very pH dependent, and theoretically the rate increases by a factor 100, when the pH increases one unit. Below pH values of 7 chemical oxydation of ferrous iron is of little importance.

The ferrous iron can also be oxidized biochemically. At low pH levels (<4) it is the same bacteria (Thiobacillus) that participate in all the oxydation steps from pyrite to ochre. At higher pH levels (>5) other bac- teria e.g. Leptotrix and Gallionella can oxidize ferrous iron, but their oxydation potential is not sufficiently studied.

It is clear from processes (1) to (4) that oxydation of pyrite is acid producing. However, the farmland in Denmark normally contains considerable amounts of basics, so that oxydation of pyrite seldom results in such a pronounded decrease in pH that the biochemical and the quick chemical oxyda- tion (4) will be of any importance. The drainage water from these soils is therefore characterized by a high concentration of ferrous iron and being slightly acid (pH 5.0-6.5).

280

High concentrations of ferrous iron in drainage water are not necessarily a result of new drainage activities. They can be due to ferrous groundwater from previous oxydation of pyrite or from other soils in the catchment area. Apart from weathering of pyrite, iron bearing clay minerals can contribute to high ferrous iron concentrations.

3. BIOLOGICAL EFFECTS

Discharge of high concentrations of ferrous iron, hydrogen ions (low pH) and ochre have serious impacts on stream ecology. It effects plants as well as invertebrates and fish. Suspended ochre increases the turbidity of the water, which results in lower primary production by macrophytes and algae. Ferrous iron precipitates on aquatic plants, pebbles and gravel, resulting in changed substrate conditions for fauna and epiphytes. Sedimentated ochre covers the spawning grounds of salmonoids and no oxygen will reach the eggs and alevins so they will die before they leave the spawning ground.

A very serious effect is that ferrous iron also precipitates on the respiratory organs of invertebrates and on fish gills, due to their slight alkaline reaction. This inevitably results in suffocation. Apart from this, ferrous iron has a direct toxic effect on fish and invertebrates.

When the pyrite containing soils do not contain sufficient bases, e.g. sandy soils in the western part of Jutland, leaching of ferrous iron is followed by high concentrations of hydrogen ions. The resulting low pH is toxic and no plants, fish and invertebrates are found at pH values below 4.0 to 4.5. The low pH may also cause higher concentrations of other substances, e.g. Al, Zn and Mn of which Al is very toxic.

Ferrous iron is toxic at very low concentrations. Some invertebrates are sensitive to concentrations as low as 0.2 mg/l, e.g. larvae of some beetles, and streams with concentrations of more than 0.5 mg/l will not contain invertebrate species normally found in otherwise clean waters (Skriver, 1984).

Low concentrations of ferrous iron also have a pronounced effect on salmonoids. At a concentration of 0.5 mg/l a reduced number of trout eggs is found, many alevins do not survive, and the populations of prey are altered. At pH values lower than 6 the reproduction of trout populations is uncertain at ferrous concentration of more than 0.5 mg/l even for stocked populations. If pH is somewhat higher (6.0-7.0) the survival of stocked populations is uncertain at concentrations of 1.0-1.5 mg/l (Geertz-Hansen et al., 1984). The cyprinids (bream, roach, eel, etc.) are more resistant to ferrous iron than salmonoids. When pH is between 6.0 and 9.0 the fry is affected at concentrations of 1 mg/l, and the populations cannot survive at concentrations above 2 mg/l (Geertz-Hansen et al., 1984).

It can be concluded that stream pollution with ochre results in a smaller number of species and fewer individuals, in contradiction to pollution with organic matter, which often results in less species but many individuals (Hunding, 1984).

4. DANISH OCHRE ABATEMENT LEGISLATION

In order to investigate the problems caused by ferrous iron leaching from pyrite-rich soils and to find methods of preventing pollution of streams with ochre, the Danish government passed a bill to make a temporary experimental arrangement possible from 1981 to 1984. The aims of the experimental arrangement was primarily:

- to map the areas where drainage may give rise to pollution with ochre;
- to test different methods of ochre abatement;
- to find limits for iron concentration below which fauna and flora population do not suffer.

The experimental arrangement was in force in five counties of Jutland (Northern Jutland, Viborg, Ringkjøbing, Ribe and Southern Jutland). It was administratively connected to the Drainage and Irrigation Subsidies Act. When a farmer applied for a government's subsidy on drainage, an investigation of the pyrite content of the soil had to be made. When the pyrite content of the soil was higher than 0.5%, the local county decided whether an ochre abatement plant was necessary in order to protect the recipient of drainage water. When the Ministry of the Environment approved the ochre abatement plant if made the necessary funds available for construction of the plant and the running costs. The plants were designed by the Danish Land Development Service. The farmer was responsible for running the plant and the local county controlled the iron removal efficiency.

This paper primarily deals with methods of ochre abatement.

5. METHODS OF OCHRE ABATEMENT

The methods of ochre abatement can be divided into two main types:

- leaching of ferrous iron is reduced by adding basics to the pyrite-rich soil prior to or just after drainage;

- the ferrous drainage water is treated before it is passed into the
streams.

Under the above mentioned three years experimental arrangement 20 ochre
abatement plants were built. Together with previously investigated ochre
abatement plants data are available from 21 plants (Christensen and Olesen,
1984). In addition experiments have been made to investigate ochre abatement
methods in streams (Waagepetersen et al., 1984).

5.1. Treatment of pyrite-rich soils

The aim of soil treatment by basics is to precipitate the iron in the
subsoil before it reaches the stream. The principle of the methods is to
add soft lime stone ($CaCO_3$) to the pyrite-rich subsoil. This will theoretically restrain the oxydation of pyrite and precipitate the dissolved iron
compounds as ochre in the soil in the following way:

$$Fe^{2+} + 1/4 \ O_2 + 2 \ H^+ + 2 \ SO_4^{2-} + 2 \ CaCO_3 + 1/2 \ H_2O \rightarrow$$

$$Fe(OH)_3 + 2 \ SO_4^{2-} + 2 \ Ca^{2+} + 2 \ CO_2 \tag{5}$$

The theoretical amount of lime is the amount of pyrite-iron multiplied by
2.2.

The incorporation of limestone into the soil is either done by deep
rotovating or by deep ploughing. Deep rotovating is performed in the following way. The soil is ploughed to a depth of about 40 cm. The limestone is
then added to the bottom of the furrow and mixed in the subsoil by rotovating to a depth of up to 90 cm. Deep ploughing is done by ploughing to a
depth of 80-90 cm with a special plough with a small mouldboard. The limestone is then added to the sides and the bottom of the furrow. The aim of
the methods is to retain most of the top soil near the soil surface, so that
no acid and nutrient-low subsoil is brought to the surface. Amounts of 20
to 400 tonnes of limestone per ha have been used in the experiments.

5.2. Treatment of ferrous drainage water

As stated in Section 2 drainage water from Danish pyrite containing
soils is often slightly acid with a high concentration of ferrous iron.
Both the chemical and biochemical oxydation of ferrous iron proceed at a
very low rate at pH-values below 7, so in order to purify the water one has
to increase either the chemical or the biochemical oxydation.

5.2.1. Oxydation ponds

In oxydation and sedimentation ponds the drainage water is given a long retention time in order to increase the ferrous iron oxydation and sedimentation of precipitated iron particles (ochre). The chemical oxydation of ferrous iron proceeds at a rate depending upon the retention time of the water in the pond, the temperature, the oxygen concentration and especially the pH-value of the water.

5.2.2. Biochemical ochre abatement

The biochemical oxydation of ferrous iron can be increased by creating large substrate surfaces accessible to ferrous iron oxidizing bacteries. These surfaces can be created in the following ways:

- cobble and pebble lined ponds with a water depth of 5-10 cm;
- weed covered oxydation ponds with a water depth of 5-10 cm;
- weed covered oxydation ponds with a water depth of about 25 cm;
- an active sludge system in which the precipitated ochre acted as the substrate surface;
- a biocontactor, consisting of a rotating cylinder containing plastic balls (bio-drum);
- a biocontactor, consisting of a number of drainage pipes through which the water is recirculated several times.

Only weed covered ponds have been put into practice. The other methods were tested experimentally in small scale models only.

5.2.3. Chemical ochre abatement

The chemical oxydation of ferrous iron can be increased by treating the water with calcium hydroxide ($Ca(OH)_2$) so that the pH of the water is increased to 7.0-7.5. In this pH-range the ferrous iron is rapidly oxidized to ferric hydroxide and precipitates as ochre according to:

$$Fe^{2+} + 2 H^+ + 2 SO_4^{2-} + 1/4 O_2 + 1/2 H_2O + 2 Ca(OH)_2 \rightarrow$$

$$Fe(OH)_3 + 2 Ca^{2+} + 2 H_2O + 2 SO_4^{2-} \tag{6}$$

After treatment with calcium hydroxide the ochre was precipitated in a pond. The same reactions take place when quick lime is used as the calcium hydroxide source. The consumption of quick lime is 2 to 3 mg per mg of ferrous iron depending upon the initial pH of the drainage water.

5.3. The efficiency of different ochre abatement methods

The results of some methods of ochre abatement are shown in Tables 1 and 2. The most efficient method is by treating the ferrous water with quick lime. On the average 89% of the ferrous iron was oxidized in the plants. When the pH of the drainage water was increased to at least 7.3, 97% of the ferrous iron was removed. At an average retention time of 5 hours in the sedimentation ponds only 66% of the total iron content was removed (Christensen and Olesen, 1984). Sedimentation can be improved by increasing the retention time.

In addition Waagepetersen et al. (1984) found that 95-100% of the ferrous iron of stream water could be oxidized when quick lime was added directly to the streams and the pH was increased to about 7.6. The sedimentation of the precipitated iron was, however, very slow, but could be increased by moderate stirring of the water or by adding floccuating agents.

Simple sedimentation ponds were not very efficient. On an average 36% of the ferrous iron was oxidized, and 32% of the total iron was sedimentated as ochre (Christensen and Olesen, 1984). The low iron removal is probably due to the low pH value of the drainage water which usually was between 5.5 and 6.5 because in this pH-range both chemical and biochemical oxydation proceed at a very low rate.

The iron removal efficiencies of cobble and pebble lined ponds and weed covered oxydation ponds with water depths of 5-10 cm were also very low. On an average only 36% of the iron was oxidized and sedimentated. The surfaces

TABLE 1   Purification of ferrous drainage water. Average figures (after Christensen and Olesen, 1984)

Method	Number of plants	Ferrous iron removal (%)	
		Oxydation	Sedimentation
Quick lime	4	97* 89	66
Oxydation and sedimentation ponds	8	– 36	32
Cobble and pebble lined ponds and weed covered ponds, 5-10 cm of water	2	– 36	36
Weed covered ponds, about 25 cm of water	1	– 78	78

*pH in outlet higher than 7.3

TABLE 2   Treatment of pyrite-rich soils. Average iron removal (after Christensen and Olesen, 1984)

Method	Number of plants	Precipitated iron (%)		Efficiency %
		Experimental area	Control area	
Deep rotovating	4	69	42	47
Deep ploughing	1	54	35	30

of the cobbles and pebbles were not sufficient to increase the biochemical oxydation rate. In the weed covered ponds small brooks were created so that the water did not come into contact with the active surfaces (Olesen et al., 1979).

The biochemical oxydation could, however, be improved considerably by increasing the water depth in weed covered oxydation ponds to about 25 cm. The average iron removal was then 78% (Olesen et al., 1979).

The experiments in which the biochemical active surfaces consisted of artificial materials also showed promising results. Water from a lignite field with a low pH-value (3.6-4.1) and a high concentration of ferrous iron (300 mg/l) was treated in a bio-drum. This resulted in an increase of the rate of ferrous iron oxydation by a factor 10 000 to 100 000 as compared to the theoretical chemical rate (Sode et al., 1984). The bio-drum was also tested on farm land drainage water with a higher pH (6.0) and a much lower concentration of ferrous iron (18 mg/l). In this case the rate of ferrous iron oxydation increased by a factor 10 as compared to the theoretical chemical rate (Nielsen et al., 1984). The method is, however, only tested in small model plants, and this makes it difficult to predict the iron removal in a full scale plant.

The limestone treatment of pyrite rich subsoils did not result in considerable iron precipitation in the soil neither did deep rotovating or deep ploughing (Table 2). The average precipitation of oxydized pyrite-iron in the subsoils was 47% when the soil was rotovated deeply. The iron precipitation was only 30% when the soil was deep-ploughed (Christensen and Olesen, 1984). The better effect of deep rotovating as compared to deep ploughing is probably due to a better mixing of limestone and soil.

6. OCHRE ABATEMENT ECONOMY

During the temporary experimental arrangement from 1981 to 1984 20 ochre abatement plants were constructed. The costs of construction varied consid-

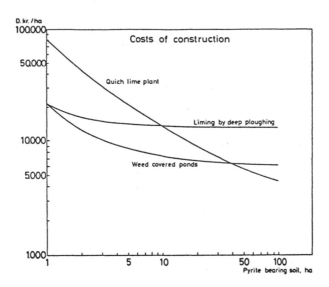

Fig. 1   Estimated costs of construction of three selected ochre abatement methods (Christensen, 1984b)

erably due to different technical solutions, treatment of drainage water
from areas between 0.5 to 35 ha in size and different pyrite contents of the
soils. The costs of construction varied between D.kr. 7000 and 380 000 (US $
1 = D.kr. 11). The most expensive solutions are treatment of the ferrous
drainage water with quick lime and limestone treatment of the pyrite-rich
soil by deep rotovating. A combination of these two methods was used in the
most expensive plant. The average costs of construction of the ochre abatement plants (D.kr. 15 000 per ha) have been in the same order of magnitude
as the costs of construction of drainage schemes (Christensen, 1984a). All
the plants have been constructed as part of the experimental arrangement and
hopefully it is possible to lower the costs when ochre abatement plant projects become a routine.

Figure 1 shows the generalized figures of construction costs for three
ochre abatement methods. For the quick lime plants the costs of construction
per unit area of pyrite-rich soil generally vary inversely with the size of
the area. With liming by deep ploughing the costs of ochre abatement is almost directly proportional to the size of the pyrite-rich area.

The calcium hydroxide method is the most efficient one, but it is also
very expensive in case of drainage of relatively small areas (5-10 ha). In
this instance weed covered oxydation ponds are relatively cheap, especially
if a slightly lower iron removal efficiency can be accepted. The quick lime

287

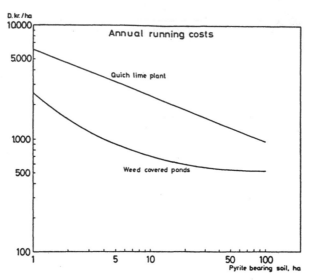

Fig. 2   Estimated running costs per year of two selected ochre abatement methods (Christensen, 1984b)

TABLE 3   Grouping of tested ochre abatement methods according to their
iron removal efficiency

Group	Average iron removal	Method
I	>95%	quick lime
II	>70%	biochemical methods e.g. weed cover-ed ponds, 25 cm of water
III	<50%	oxydation and sedimentation ponds, weed covered ponds, 5-10 cm of water, limestone treatment of the soil by deep rotovating and deep ploughing

treatment method is the cheapest one when the area exceeds about 50 ha.

Data for running costs of the 20 ochre abatement plants are still very
scarce. Figure 2 gives the estimated running costs for the same three ochre
abatement methods given in Figure 1. Limestone treatment of subsoil by deep
ploughing gives no working expenses. The running costs of the quick lime
plants are much higher than the ones for weed covered oxydation ponds. On
the other hand, the quick lime method is much more reliable, especially at
low temperatures.

7. CONCLUSION

As regard the future use the different ochre abatement methods can be

288

divided into three groups according to their iron removal efficiency (Table 3). It can be concluded that for the time being treatment of ferrous drainage water with quick lime is the only efficient and reliable method when a high protection of the recipient is necessary. Some of the biochemical ferrous iron oxydation methods also have a relatively high iron removal efficiency. The methods are not yet fully developed, but their efficiency can probably be increased to the same high level as the quick lime method. The iron removal of all the other ochre abatement methods tested was very low (30-50%), and their use is normally not recommended.

REFERENCES

Christensen, L.B. 1984a. Okkerrensning - metoder, rensningseffekt, drift og økonomi (Ochre abatement - methods, efficiency, workings and economy). Vand og Miljø 3: 11-15.

Christensen, L.B. 1984b. Okkerrensning (Ochre abatement). Hedeselskabets tidsskrift, Vaekst. 105.5: 18-21.

Christensen, L.B. and Olesen, S.E. 1984. Forsøg med okkerrensning ved landbrugsmaessige draeninger (Ochre abatement experiments in connection to farm land drainage). Bilag 3 til okkerredegørelsen, Miljøstyrelsen. 186 pp.

Geertz-Hansen, P., Nielsen, G. and Rasmussen, G. 1984. Fiskeribiologiske okkerundersøgelser (Investigations of fish biology in connection to ochre pollution). Bilag 8 til okkerredegørelsen, Miljøstyrelsen. 169 pp.

Hunding, C. 1984. Okker kan true vore vandløb (Ochre is a threat to our streams). Hedeselskabets tidsskrift, Vaekst. 105.5: 12-13.

Nielsen, S., Jacobsen, J., Waagepetersen, J. and Olesen, S.E. 1984. Afprøvning af metoder til biologisk iltning af ferrojern i vand fra Hvidmosen (Experiments on biochemical oxydation of ferrous iron in drainage water from the Hvidmosen). Bilag 16 til okkerredegørelsen, Miljøstyrelsen. 23 pp.

Olesen, S.E., Larsen, V. and Hansen, K.O. 1979. Forsøg med rensning af vand for indhold af jern (Experiments on purification of iron-rich drainage water). Beretning nr. 22, Det danske Hedeselskab. 84 pp.

Skriver, J. 1984. Okkers indvirkning på invertebratfaunaen i midt- og vestjyske hedeslettevandløb (The effect of ochre on invertebrates in streams of central and western Jutland). Bilag 8 til okkerredegørelsen, Miljøstyrelsen. 66 pp.

Sode, A., Jacobsen, J. and Nielsen, S. 1984. Biotromleforsøg. Rensning af jernholdigt afløbsvand med lavt pH (Experiments with bio-drums. Purification of ferrous run-off water with low pH). Bilag 5 til okkerredegørelsen, Miljøstyrelsen, 47 pp.

Waagepetersen, J., Grant, R.O. and Olesen, S.E. 1984. Forsøg med okkerrensning i vandløb - Hvirlå og Grundel baek (Experiments on iron removal in streams - river Hvirlå and Grundel brook). Bilag 6 til okkerredegørelsen, Miljøstyrelsen. 93 pp.

Nitrogen balance in crop production and groundwater quality

H.C.ASLYNG
Hydrotechnical Laboratory, RVAU, Copenhagen, Denmark

ABSTRACT

An inorganic NITrogen balance in CROp production Simulation model NITCROS has been developed and field experiments were carried out in 1982-1984. The investigation covers several agricultural crops and were done at Karup (56°16' N, 9°9' E) with an average annual precipitation of 775 mm on coarse sand (soil 1), and at Tåstrup (55°40' N, 12°18' E) with an average precipitation of 600 mm and on clay with fine sand (soil 6). Soils 1 and 6 have 5 and 8 t N/ha in organic matter, respectively, to a soil depth of 50 cm and an average effective available water root zone capacity of 60 and 170 mm.

Data for soils 1 and 6 in kg N/ha as annual average (with great annual variation) were found to be for mineralization 65 and 120, for denitrification 3 and 20, for uptake in harvest 170 and 200, for leaching 50 and 40, and for spring storage 10 and 50, respectively. Winter and late growing crops reduced the leaching by around 20 kg N/ha. Nitrogen fertilizer should be applied relatively late to sandy soils and early to clay soils. Fertilization with inorganic fertilizer had only little influence on leaching losses. Nitrate is reduced in clay underground, which protect the groundwater, whereas with a sandy underground the nitrate concentration in groundwater increases. The nitrate concentration in groundwater is increasing, especially in sandy regions where large amounts of manure are applied.

1. INTRODUCTION

The NITrogen balance in CROp production Simulation model NITCROS is described by Hansen and Aslyng (1984) and Aslyng and Hansen (1985).

The following components of inorganic nitrogen are included in the balance:

$$N_a + N_f + N_m - N_d - N_1 - N_u = \Delta N_i$$

where N_a = atmospheric fall-out and soil microbial fixation

N_f = inorganic fertilizer

N_m = mineralization

N_d = denitrification

N_1 = leaching

N_u = uptake by plants

ΔN_i = change in soil inorganic nitrogen

Application of manure or organic fertilizer is not included. The model can be considered as an extension of the model WATCROS (Aslyng and Hansen, 1982, 1985). Simulation with NITCROS during the growth period requires both WATCROS input variables and variables generated by WATCROS. To simulate during the winter period NITCROS requires daily values of air temperature, precipitation and potential evapotranspiration. In addition NITCROS requires a number of soil and crop parameters. The required variables and parameters are stated in the description of the submodels for components of the inorganic balance. The inorganic nitrogen in the soil is distributed over individual soil layers, which is described in the leaching submodel.

2. ATMOSPHERIC FALL-OUT AND SOIL MICROBIAL FIXATION

The atmospheric fall-out and soil microbial fixation, N_a, are assumed to be constant independent of season and precipitation, N_a = 0.05 kg N·ha^{-1}· d^{-1}. The annual amount is 18 kg N/ha available N of which 12 kg come from precipitation and 6 kg from microbiological fixation.

3. INORGANIC FERTILIZER

Broadcasted fertilizer must be dissolved before it becomes available to plants and microorganisms. The fertilizer is assumed to be dissolved if one of the following three criteria is fulfilled:

- the uppermost soil layer contains at least 80% of the capacity of available water;
- the precipitation sum over the present and the previous 2 days is at least 7 mm;
- the precipitation sum over the present and the previous 6 days is at least 12 mm.

4. MINERALIZATION

The mineralization - immobilization turnover constitutes a very complex system. The mineralization or, to be more precise, the net mineralization model takes only into consideration the most important facts. It is a basic assumption that the soil is well aerated and has a pH within the range 5-8.

The daily net mineralization is predicted by use of a rate constant, the soil organic nitrogen content (5-9 t N/ha in the upper 50 cm), the soil temperature and the soil water effect. A rate constant of $0.7.10^{-4} \cdot d^{-1}$ is used. To represent the soil temperature the temperature at 10-15 cm depth is taken. If the temperature is not recorded it is estimated from the equation:

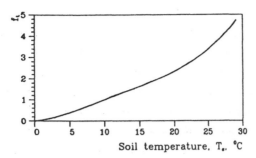

Fig. 1   The effect of soil temperature on mineralization

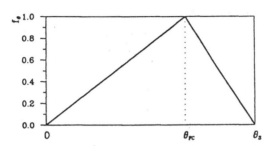

Fig. 2   The effect of soil water content, θ, on mineralization

$$T_s = 0.66 + 0.93\ \overline{T_a}$$

where $\overline{T_a}$ is average air temperature over the present and previous 6 days.
Temperature and soil water effects are illustrated in Figures 1 and 2.

5. DENITRIFICATION

The denitrification model is extremely simple. It comprises a denitrification pulse and a diffuse denitrification term. A denitrification pulse is often recorded just after the applied fertilizer is dissolved. This pulse is assumed to be proportional to the applied amount of fertilizer N_f and a proportionality factor (from zero to 0.1) depending on the soil type.

In addition to the denitrification pulse a diffuse denitrification is assumed to occur when the soil is sufficiently warm (criterium I), sufficiently moist (criterium II) and the NO_3^--concentration is sufficiently high (criterium III). The adopted criteria are:

I   soil temperature at least 5°C
II   soil water at least 90% of available capacity
III   at least 0.6 kg NO_3-N per cm soil/ha

293

6. LEACHING

In order to predict leaching of inorganic nitrogen a slightly modified version of the leaching model proposed by Addiscott (1977) is applied. This model has been especially developed for leaching in structured soils. It is a layer model in which transport of nitrogen between layers takes place as solute in percolating water as mass flow. Diffusion is not considered.

As a working hypothesis the soil solution is partitioned into a mobile (MW) and an immobile (RW) fraction. The latter represents water held in 'dead end' pores and very small pores which are assumed as not contributing to mass flow. Accordingly only the MW part of the solution is displaced during water movements. Equilibrium is assumed to be established between MW and RW after each flow event. The time step of the model is one day and a flow event is a day on which flow between layers occurs. The model permits displacement of the mobile solution through an infinite number of layers when large flows occur.

The MW and RW values in each layer are important parameters in the Addiscott model. The allocation of values to MW and RW is essentially arbitrary because of the nature of the working hypothesis. Addiscott proposed the upper limit of the MW to be pF 1.7 (field capacity) and the lower limit of the RW to be halfway between pF 4.2 (wilting point) and dryness. The partition between mobile and retained water is taken at pF 3.3.

The omission of some strongly held water is allowed for the fact that diffusion in this water will be very slow and that anion exclusion occurs. Both anion exclusion and strongly held water are related to the colloidal content of the soil. The amount of water assumed to be free from mineral nitrogen represents only a few layers of water molecules on the surface of the colloidal particles.

In our version of the Addiscott model the uppermost layer is treated separately. The underlying layers, each of the thickness of 5 cm, are treated as proposed by Addiscott (1977). In spring before onset of growth the uppermost layer has the thickness of 25 cm, but when the roots penetrate into lower layers, these layers are included in the uppermost layer.

After the crop harvest the soil profile is divided into layers as before the onset of growth, while at that moment a uniform distribution of water and nitrogen is assumed. Mineralization and uptake of water and nitrogen are assumed to take place in the uppermost layer. Nitrogen is assumed to be at equilibrium and leaching is assumed to take place with the equilibrium concentration and instantaneously for layers below the uppermost layer, when field capacity is reached (as in the Addiscott model). The water balance is run as a book-keeping system with input and output on a daily basis.

7. UPTAKE BY PLANTS

In the simulation model NITCROS the crop is characterized by the accumulated dry matter, the effective root depth and the age of the crop, all of which are simulated by WATCROS. In addition the crop is characterized by the N-concentration. The soil is characterized by its content of inorganic nitrogen and a minimum content below which the plants cannot extract nitrogen from the soil. So it is assumed that either the nitrogen demand of the crop or the nitrogen content of the soil limits the uptake. The nitrogen demand is estimated from a concentration in the crop (top and root) and from the accumulated dry matter in the crop. The concentration is assumed to be a function of the crop species and the age of the crop (Figure 3).

For cereals the nitrogen uptake ceases 10 days before the crop is ripe for harvesting. The root zone minimum content of inorganic nitrogen not utilized (left as unavailable) by a crop is a soil type characteristic value (Lindén, 1983; Østergaard et al., 1983; Andersen and Jensen, 1983; Aslyng and Hansen, 1985). We have assumed for coarse sand and for clay with fine sand 5 and 20 kg N/ha as the minimum root zone content, respectively, for fertilized crops.

When the production is nitrogen limited then respiration losses are recalculated to be in agreement with the actual production in the same way as respiration is calculated by WATCROS. The model is operated as a book-keeping system which keeps account of the nitrogen status of the crop. In the case of grass an amount of nitrogen is predicted to be removed with the dry matter in each cutting. In case of other crops which are harvested only

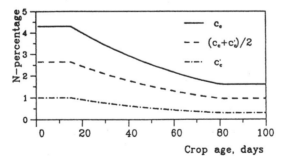

Fig. 3   Crop nitrogen concentration regulating nitrogen uptake and crop production of spring barley. C_c upper limit for nitrogen uptake, C_c' lower limit for production and $(C_c+C_c')/2$ lower limit for maximum production

once, the total crop uptake is predicted. The harvested nitrogen is estimated from the total uptake minus the values for the residue.

Winter crops already contain some nitrogen at the onset of the growth in spring. For winter wheat and winter barley not fertilized in the autumn we have adopted this quantity to be 20 kg N/ha and for new established (undersown) one-year grass and two-year grass 25 and 50 kg N/ha, respectively.

8. EXPERIMENTAL AND PREDICTED DATA

Determined and predicted nitrogen in harvested yield increases almost linearly during the first 5 to 6 weeks of the growth period (Figure 4). The dates at the end of this first period and in brackets the harvested quantity of kg N/ha, are about as follows: winter wheat 1.6 (190), summer barley 15.6 (110), summer rape 20.6 (210) and fodder beets 15.7 (200). For grass there is a linear increase until 1.8 (400) but this is depending on the split application of nitrogen.

The period of linear increase in nitrogen yield is the period during which applied fertilizer nitrogen and the spring storage and mineralized nitrogen is utilized. The quantities in brackets are close to the amounts of applied fertilizer nitrogen.

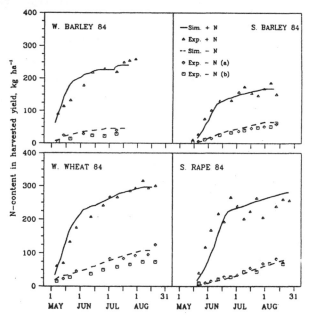

Fig. 4   Nitrogen yield. In case (a) the actual and in case (b) also the previous crop was unfertilized. In cases (a) and (b) yields differ for winter but not for spring crops. The simulated yield of rape comprises also N in the shedded leaves

After the first period of 5 to 6 weeks the increase in nitrogen yield represents mineralized nitrogen apart from small quantities extracted from deeper layers by an increased root depth. A considerable uptake after anthesis is required for large grain yields. This amount of nitrogen may be mineralized in case of clay soil but not in case of sandy soils. With low mineralization a split and late nitrogen application is needed for large yields.

The major influence of applied nitrogen on yield is found in the number of ears or pods per m^2. An additional small influence is observed in the number of grain/ear and seed/pod. The strong influence on the number of ears or pods per m^2 suggests that a relatively early nitrogen application is necessary.

A single early nitrogen application is recommended to soils in which the root depth is large and where the precipitation is low in the early part of the growth period. That means application about 2 weeks before the onset of spring growth for winter crops and at seed drilling for spring crops. Nitrate transport into the upper soil promotes root development which contributes to total water and nitrogen uptake by the crops. Under conditions of sandy soils and 'large' precipitation, nitrogen application is recommended to be 2 to 3 weeks later (Table 1) and also split application may be recommended.

With the recommended time for nitrogen application the annual nitrogen losses by leaching will depend on soil type and on mineralization and net precipitation after harvest. Early ripening increases and winter crops reduces the losses (Table 2). The predicted average annual nitrogen loss by leaching in Denmark amounts to about 40 and 50 kg N/ha from clay and sandy soils supplied with inorganic fertilizer. Organic fertilizers increase the leaching loss.

TABLE 1 Nitrogen leaching and harvest depending upon time and quantity of nitrogen application to summer barley at Karup. Annual average 1968-1976

| Nitrogen, kg N/ha | | Leaching | Harvest | Relative harvest |
At seed drilling	At onset growth	kg N/ha	t/ha	
140	–	68	8.4	92
70	70	61	8.8	97
–	140	53	9.0	99
–	120	52	8.8	97
–	70	51	6.4	70

TABLE 2   Predicted nitrogen turnover (kg/ha), total relative harvest (percent of water limited) and N-fertilizer application (kg/ha) to different crops. Annual average (1 April – 31 March) 1966-1984 at Tåstrup and 1968-1977 at Karup

Location and crop	Fertil-izer*	Mineral-ization	Denitrifi-cation	Leach-ing	Crop uptake	Spring storage	Relative harvest
Tåstrup							
Winter wheat	170	110	20	32	246	37	98
Summer barley	110	120	15	47	186	53	96
Summer rape	170	118	21	58	257	55	91
Fodder beet	170	127	20	19	276	28	99
Grass	363	113	38	18	438	36	83
Karup							
Winter wheat	200	62	2	42	236	5	94
Summer barley	120	65	1	52	150	7	97
Summer rape	200	64	4	90	218	7	93
Fodder beet	200	70	4	46	238	7	98
Grass	267	61	2	51	293	7	87

*In addition, 18 kg N as atmospheric fall-out and microbiological fixation

In clay undergrounds the nitrate reduces to a great extent and the nitrogen excapes to the atmosphere. In sandy undergrounds the reduction is uncertain. In agreement with this it is found that the nitrate content in groundwater in Denmark is increasing with a small rate in the eastern regions with clay soils and with a much larger rate in the western regions with sandy subsoils (Figures 5 and 6). There may, however, be a considerable time lack between leaching of nitrate from the root zone and the appearance of it in deep groundwater.

Since 1974 Denmark has had a law on environmental protection. Now further regulations concerning increased protection of surface and groundwater are treated by the Parliament. As for agriculture this especially applies to the nitrogen losses to the air and by leaching from the farms and from the fields. Farm losses are from stables and deposits of manure and fodder.

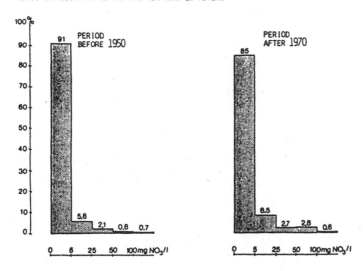

NITRATECONCENTRATIONS IN CONFINED AQUIFERS ON ZEALAND
4,400 SAMPLES FROM DEPTHS GREATER THAN 10 METERS

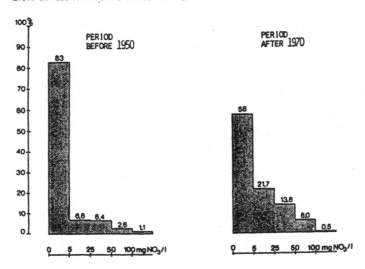

NITRATECONCENTRATIONS IN PREDAMINANTLY PHREATIC AQUIFERS IN THE WESTERN PARTS OF JUTLAND
2,800 SAMPLES FROM DEPTHS GREATER THAN 10 METERS

Fig. 5   Nitrate in groundwater (Miljøstyrelsen, 1983)

It is proposed that the farms must have storage capacities for at least
5 months production of slurry/manure and that runoff from the deposits must
be avoided or collected. It will not be permitted to apply liquid manure to
bare soil in autumn. Rules should secure a balance between number of animals

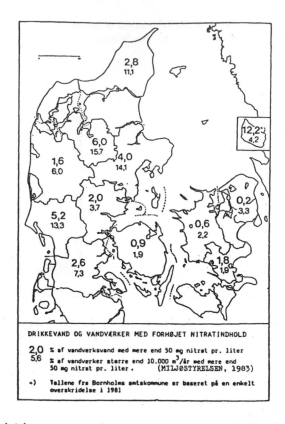

Fig. 6   Drinking water and groundwater works, %

and area of land for receiving the manure. The leaching from fields is mainly nitrogen from autumn application and mineralization after harvest. The
increased use of winter crops will reduce leaching losses somewhat. Research
and advice on efficient farming should be promoted.

REFERENCES

Addiscott, T.M. 1977. A simple computer model for leaching structured soils.
    J. Soil Sci. 28: 554-563.
Andersen, A. and Boye Jensen, M. 1983. Jordbearbejdning og efterafgrøde ved
    bygdyrkning. 2. Bygplanternes morfologiske udvikling i relation til
    kvaelstof (summary in English). Tidsskr. Planteavl 87: 217-236.
Aslyng, H.C. and Hansen, S. 1982. Water balance and crop production simula-
    tion. Model WATCROS for local and regional application. Hydrotechnical
    Laboratory. The Royal Vet. and Agric. University, Copenhagen. 200 pp.
Aslyng, H.C. and Hansen, S. 1985. Radiation, water and nitrogen balance in
    crop production. Field experiments and simulation models WATCROS and
    NITCROS. Hydrotechnical Laboratory. The Royal Vet. and Agric. Univer-
    sity, Copenhagen. 146 pp.

Hansen, S. and Aslyng, H.C. 1984. Nitrogen balance in crop production. Simulation model NITCROS, Hydrotechnical Laboratory. The Royal Vet. and Agric. University, Copenhagen. 113 pp.

Lindén, B. 1983. Movement, distribution and utilization of ammonium- and nitrate nitrogen in Swedish agricultural soils. Dissertation. Dept. Soil Science, Division Soil Fertility, Swedish University of Agric. Science, Uppsala. 39 pp.

Miljøstyrelsen. 1983. Nitrat i drikkevand og grundvand i Danmark, Copenhagen. 125 pp.

Østergaard, H.S., Hvelplund, E.K. and Rasmussen, D. 1983. Kvaelstofprognoser. Bestemmelse af optimalt kvaelstofbehov på grundlag af jordananalyser og klimamålinger før vaekstsaesonen. Landskontoret for Planteavl, Viby J. 200 pp.

Nitrate load and water management of agricultural land

P.E.RIJTEMA
Institute for Land & Water Management Research (ICW), Wageningen, Netherlands

INTRODUCTION

In the Netherlands the demand for water of good quality by agriculture, industry and municipal water supply and also for nature is steadily increasing. This calls for proper water management schemes. In order to find the optimal distribution of activities in a region one needs a weighing of the conflicting interests.

Agriculture can be characterized by an intensive use of fertilizers, a need for good drainage conditions during winter and an additional water supply in summer.

For municipal and industrial water supply the main objective is a continuous delivery of sufficient water of good quality from both surface water and groundwater resources. Proper water management in any region with competing interests implies an evaluation of the various impacts by the activities.

Special attention has been given to the effects of intensive agriculture on the nitrate load of the groundwater resources as a result of improved drainage conditions and to the effect of deep well pumping for municipal and industrial water supply on nitrate load.

EFFECT OF DRAINAGE ON NITROGEN POLLUTION

Intensification of agriculture has led to an increased fertilizer use. The potential of livestock waste to pollute surface waters and groundwater is large, particularly when plant nutrients are supplied in excess of crop requirements.

The nitrate concentrations of the groundwater in wet nature areas in sandy regions with shallow water tables is on the average 0.3 $g \cdot m^{-3}$, which gives an indication of the level of natural nitrate concentrations in deep aquifers with a groundwater table near soil surface. The nitrate concentration in the groundwater under forests with deeper groundwater tables varies from 0.2 to 22 $g \cdot m^{-3}$ N, with an average value of about 5 $g \cdot m^{-3}$ N (Oosterom and Van Schijndel, 1979).

Many data for the nitrogen loss from arable lands were given by the

Curatorium Landbouwemissie (1980). For sandy soils with a crop rotation of twice cereals, potatoes and sugar beets in four consecutive years, with a mean yearly application of 150 kg N per ha in the form of chemical fertilizer, an average concentration of 28 $g \cdot m^{-3}$ N must be expected in the shallow groundwater. When animal slurry is applied instead of fertilizer-N the concentration increases to 50 $g \cdot m^{-3}$ N. On soils where fodder maize is grown, each additional overdosing of slurry in a quantity equivalent to 50 kg N per ha increases the concentration with 4.5 $g \cdot m^{-3}$ N.

A nitrogen model for grassland has been given in a previous study (Rijtema, 1980). The average emission at a standard livestock density of 4 units per ha equals 127 to 82 kg N per ha, depending on drainage conditions.

Difficult soil and drainage conditions on part of the farm may limit the area suitable for winter application of slurry and this will result in local overdosing.

The consequences of fertilizing with regard to nitrogen pollution of ground- and surface waters depend on land use, soil type, fertilization level and on the hydrological situation. In poorly drained soils part of the precipitation is discharged by surface runoff and by transport through the shallow top layer of the root zone to the surface water. Under these conditions the residence time of both precipitation surplus and nitrogen in the soil is very short and the main part of the nitrogen present in the slurry applied in winter will be transported to the surface water during the same period.

Steenvoorden (1980) gives data of the nitrogen concentration (organic-N and NH_4^+-N) in surface runoff, when slurry is applied in winter. The amounts vary from 80 to 2 $g \cdot m^{-3}$ N, depending on the difference in time between slurry application and the occurrence of surface runoff, as well as on the surface runoff quantity.

Surface runoff does not only occur when the infiltration rate of the topsoil is less than the rainfall intensity, but also when due to poor drainage conditions the phreatic level rises to the soil surface. Table 1 shows that the quantity of surface runoff decreases with increasing drainage intensity.

The data given by Steenvoorden and Buitendijk (1980) also show that surface runoff varies from year to year depending on the distribution of rainfall. Rough estimates of the quantity of mean surface runoff based on calculations by Rijtema and Bon (1974) have been given by Rijtema (1982). This approach can be useful to transfer a hydrological classification system, as

TABLE 1 Total surface runoff (mm) during the period 1 September 1961 to 1 June 1962 for soil with a drainage depth of 70 cm and a drainage intensity of 3, 5 and 8 mm·d^{-1} and a phreatic level 20 cm below the ground surface (Steenvoorden and Buitendijk, 1980)

	Drainage intensity (mm·d^{-1})		
	3	5	8
Surface runoff (mm)	118	69	30

TABLE 2 Average hydrological data, as derived from a soil survey classification system for Pleistocene sand

Hydrological class	Water table		E	Seepage	Total drainage	Surface runoff
	mean min.	mean max.				
	cm –surface		mm	mm·d^{-1}	mm·a^{-1}	
1	0	60	530	+0.8	542	163
2	0	70	510	+0.5	453	124
3	20	90	490	+0.3	400	86
4	40	90	500	+0.4	426	53
5	30	120	447	−0.4	187	−
6	60	140	422	−0.9	30	−
7	90	160	415	−1.0	−	−

TABLE 3 Estimated nitrogen load of surface waters in relation to the basis of local drainage and mean surface runoff

Local drainage basis m –soil surface	Surface runoff mm·a^{-1}	Nitrogen load kg·ha^{-1} N
0.2	163	185
0.2	130	170
0.4	90	145
0.5	55	110
0.6	15	60
≥0.7	0	45

used in soil surveys into estimated values of surface runoff. The results of this type of calculations for Pleistocene sands are given in Table 2.

Rijtema (1982) estimated from these data the mean nitrogen load to surface water, using the local drainage basis and the surface runoff as hydrological parameters. These results are given in Table 3.

The main objective of improving local drainage is to lower the water table in winter and during extremely wet periods in other seasons. The main effect will be a reduction in the quantity of surface runoff. Well-drained soils generally have a deep percolation of the precipitation surplus and residence times in the soil of 5 years and more are not unusual. When analyzing the consequences for water quality management one has to deal, apart from the transport of matter under influence of the hydrological situation, with aerobic and anaerobic biodegradation, nitrogen immobilization, mineralization and denitrification.

It appears that on arable lands only 50% of the available mineral nitrogen is used for crop production, while the gross uptake for grasslands is 70 percent. The total loss of mineral nitrogen from the root zone equals the combined losses due to denitrification and leaching. Practical experience on grassland has shown that nitrate below a depth of 120, 110 and 80 cm in sand, clay and peat soils, respectively, can be considered as being lost for crop growth.

Leaching of nitrogen occurs for nearly 100 percent in the form of NO_3^-. In very deeply drained soils about half of the non-used nitrogen leaches to the groundwater. A sufficient aeration of the unsaturated zone and a limiting quantity of organic compounds in the deeper layers restrict denitrification, as this process requires anaerobic conditions and organic compounds for energy supply. It must be expected that denitrification increases with decreasing depth of drainage and drainage intensity as a result of wetter conditions in the topsoil. Data given by Steenvoorden (1983) show that the leaching of nitrate to the groundwater must be reduced by a coefficient that depends on the mean height of the winter water table. The relation between this reduction factor and the mean winter water table for Pleistocene sands is given in Table 4.

The total nitrate pollution of groundwater aquifers is not the result of human activity on a single field, but it depends on the regional distribution of land use, soil type, fertilization level and drainage conditions. As

TABLE 4   The reduction coefficients by which the nitrate leaching from deeply drained soils must be multiplied as a function of the mean depth of the groundwater table in winter

Mean water table depth in m -surface	0.0	0.2	0.3	0.4	0.6	0.9	1.0
Reduction factor	0.04	0.10	0.15	0.22	0.41	0.73	1.0

a consequence, studies of aquifer pollution by agricultural activities must
be regional.

EFFECTS OF DEEP WELL PUMPING ON GROUNDWATER POLLUTION

Where in humid areas the ground surface has only relatively small differences in elevation and the transmissivity of the aquifer has an appreciable value, excess precipitation is mainly carried-off by groundwater flow to
a system of rather closely spaced drains of different size and level. The
depth of the groundwater table and the rate of discharge are variable due to
the seasonal fluctuations of evaporation and the irregular variations in
precipitation. Under these conditions river discharges generally will be
large in winter and small in summer.

The topography of the country more or less prohibits the construction of
large artificial reservoirs to store winter discharge for municipal and industrial water supply. This forces water supply companies to exploit groundwater resources for this purpose.

Deep well pumping of groundwater from thick phreatic or semi-confined
aquifers will cause a drawdown of the phreatic level, particularly in the
case of phreatic aquifers. Primarily this results into a smaller discharge
of water by surface drains. Near the centre of the catchment area of a well,
the drainage system does not operate anymore due to the drawdown of the
groundwater table, whereas the drainage system still works at greater distances.

In those situations, where, without pumping, the depth of the phreatic
level during summer was rather small, a reduction in evaporation during
periods of drought has to be taken into account under conditions of deepwell
pumping. As the growth of crops depends on available soil water, which in
turn depends on the depth of the phreatic level shows that groundwater extraction can cause reductions in yield of agricultural crops and therefore
also in nitrogen uptake by the plant.

Ernst (1971) gives a model formulation for the calculation of the mean
drawdown of the phreatic level in relation to the distance from a deep well,
as a function of the hydrological parameters of the catchment. The effect of
the mean drawdown of the phreatic level on the mean highest winter groundwater table and the mean deepest summer groundwater table has been analyzed by
Rijtema and Bon (1974).

In the ECE-guideline for water to be used for municipal water supply,
the directive for the maximum acceptable nitrate concentration is 50 $g \cdot m^{-3}$

(11.3 $g \cdot m^{-3}$ N), while a recommended level of less than 25 $g \cdot m^{-3}$ NO_3 (5.6 $g \cdot m^{-3}$ N) is given. Data, both from analyses of extracted groundwater in a number of deep wells and from locally or regionally executed groundwater quality research show that groundwater in many locations has been polluted by nitrate.

For the calculation of agricultural impacts on regional groundwater quality the model NIMWAG (Rijtema and Hoeymakers, 1985) can be used. This model has three main components:

- a hydrological part based on the formulation given by Ernst (1971) for the calculation of the distribution of the groundwater recharge, the pattern of drawdown of the phreatic level and the calculation of the residence time of the infiltrating precipitation surplus before reaching the deep well;
- a nitrogen fertilization part for the calculation of the fertilizer level in the region, as a function of land use, slurry production and introduced restrictions in slurry application because of environmental conditions;
- a part for the calculation of the nitrate flow towards the aquifer as influenced by denitrification and transport processes.

Different types of land use can be introduced in the region under consideration. The model distinguishes 10 different types of land use. For a study on national scale of the agricultural impacts in catchment areas of 166 deep wells for municipal water supply only six types of land use were used. These types were:

- intensively used grassland, with day and night grazing
- arable land with a crop rotation of cereals, potatoes, sugar beets
- arable land with fodder maize
- horticulture
- forests
- urban areas

For the calculation of the effect of preventive measures on the nitrate load of the aquifer, various restrictions in slurry and fertilizer application for each type of land use were introduced. The restrictions were:

- no excess application of slurry (no dumping);
- restriction 1 + no land spreading of slurry in autumn and winter;

TABLE 5 The effect of restrictions in slurry application on the mean nitrogen concentration of water infiltrating into the aquifer for two different catchments for the situation without and with deep well pumping

Fertilization	Nitrogen concentration recharge water ($g \cdot m^{-3}$ N)			
	Catchment A		Catchment B	
	Deep well pumping		Deep well pumping	
Restriction	no	yes	no	yes
No restriction	3.0	8.7	20.0	25.8
No dumping	2.8	8.2	16.8	22.2
Spring application	2.2	6.3	13.4	17.6
75% N-requirement	1.9	5.1	11.7	14.8
50% N-requirement	1.6	3.9	9.8	12.4

- restriction 2 + suboptimal N fertilization level of 75 and 50 percent respectively.

These restrictions in fertilization were applied to various regions in the framework of regulations in a general soil protection law, as well as for limited protection areas with a residence time of less than 100 years in catchment areas for municipal water supply. Table 5 shows some typical results for two regions. These regions differ considerably in land use and hydrological conditions. Region A was formerly a wet region with intensively used grassland. It appears that without groundwater extraction application of restriction in slurry application is not necessary, as the concentration of the infiltrating water remains below the recommended level of 5.6 $g \cdot m^{-3}$ N. The drawdown of the groundwater table due to deep well pumping reduces denitrification in the area, resulting in an increased concentration. Although the maximum accepted level is not reached, it appears that the recommended boundary value only can be met with sub-optimal fertilizer. Region B is an area with mainly arable land and forests. The groundwater table, without pumping, is already very deep in the major part of the catchment. In this case the water recharging the aquifer has already concentrations exceeding the maximum allowed level for municipal water supply, except with sub-optimal fertilization at 50%. The small increase in N-concentration with deep well pumping is enough to prohibit even this situation.

In a recently published study of the Werkgroep Nitraatuitspoeling Waterwingebieden (1985) an analysis for 152 catchments has been given. The number of catchments in which difficulties can be expected in future, applying dif-

TABLE 6   Number of catchments in which the mean concentration of the infiltrating water in the aquifer exceeds the maximum accepted concentration of 11.3 $g \cdot m^{-3}$ NO_3-N for municipal water supply at different restrictions in fertilizer application

Protected area with residence time $\leqslant$100 years	Total region (outside protected area)				
	No restrictions	No dumping	Spring application	Sub-optimal fertilization	
				75%	50%
No restrictions	75	–	–	–	–
No dumping	72	66	–	–	–
Spring application	61	58	41	–	–
75% N-requirement	52	48	34	25	–
50% N-requirement	42	35	22	13	10

ferent restrictions in the use of animal slurrie and N-fertilizer is summarized in Table 6.

It appears that with the present maximum level of maximum accepted nitrate concentration of 11.3 $g \cdot m^{-3}$ N in large number of catchments difficulties must be expected.

CONCLUSIONS

It appeared that under poor drainage conditions the nitrogen load to surface waters strongly depends on surface runoff and on the concentration of organic N and NH_4^+-N in the water. Improvement of drainage conditions reduces surface runoff and the N load of organic N and NH_4^+-N. However, due to better soil aeration, there will be a reduction in denitrification and an increased discharge of nitrate. When the drop in water table is only small, the net result will be a reduction in the total N-load to surface waters.

The nitrate load of the aquifer depends strongly on the land use and the hydrological conditions in the region. It appeared that deep well pumping increased the nitrate load, due to the decline in phreatic water level, combined with a reduction in denitrification, even without changing the land use intensity.

The present maximum level of nitrate of 11.3 $g \cdot m^{-3}$ N is a real problem for the development of agricultural activities in catchments with deep wells for municipal water supply. Reduction of fertilization to a sub-optimal level of 50% does not offer a solution in all cases.

The weighing of the competing interests of agriculture and municipal and industrial water supply is not a small scale local problem. Extrapolating the results of the present analysis to the 300 pumping stations expected in future results in a total protection area of 170 000 ha, from which 105 000 ha is agricultural land.

REFERENCES

Curatorium Landbouwemissie. 1980. Rapport over emissie vanuit de landbouw (Report of emissions from agriculture, in Dutch). Ministerie van Landbouw en Visserij.

Ernst, L.F. 1971. Analysis of groundwater flow to deep wells in areas with non-linear function for the subsurface drainage. Techn. Bull. 75 Institute for Land and Water Management Research (ICW), Wageningen.

Oosterom, H.P. and Van Schijndel, J.H.W.M. 1979. De chemische samenstelling van het bovenste grondwater bij natuurlijke begroeiingen op kalkarme zandgronden (The chemical composition of the phreatic groundwater under natural plantcovers on non-calcareous sandy soils, in Dutch). Nota 1075 Institute for Land and Water Management Research (ICW), Wageningen.

Rijtema, P.E. 1980. Nitrogen emission from grassland farms - a model approach. Proc. Int. Symp. Eur. Grassland Fed. on the role of nitrogen in intensive grassland production. Wageningen, The Netherlands: 137-147. PUDOC, Wageningen.

Rijtema, P.E. 1982. Effects of regional water management on N-pollution in areas with intensive agriculture. Report 4 Institute for Land and Water Management Research (ICW), Wageningen.

Rijtema, P.E. and Bon, J. 1974. Bepaling van landbouwkundige gevolgen van grondwaterwinning met behulp van bodemkundige gegevens (Determination of agricultural consequences of groundwater extraction, using soil survey data, in Dutch). Reg. Studie 7 Institute for Land and Water Management Research, Wageningen.

Rijtema, P.E. and Hoeymakers, T.J. 1985. The nitrate conflict between municipal water supply and agriculture. Nota 1423 Institute for Land and Water Management Research, Wageningen.

Steenvoorden, J.H.A.M. 1980. Eutrophication. Research Digest 1980. Techn. Bull. 117 Institute for Land and Water Management Research, Wageningen: 121-125.

Steenvoorden, J.H.A.M. 1983. Nitraatbelasting van het grondwater in zandgebieden; denitrificatie in de ondergrond (Nitrate load of groundwater in sandy areas; denitrification in the saturated zone, in Dutch). Nota 1435 Institute for Land and Water Management Research, Wageningen.

Steenvoorden, J.H.A.M. and Buitendijk, J. 1980. Oppervlakte afvoer (Surface runoff, in Dutch). In: Waterkwaliteit in grondwaterstromingsstelsels. Rapport Comm. Hydrologisch Onderzoek TNO 5: 87-92.

Werkgroep Nitraatuitspoeling Waterwingebieden. 1985. Nitraat bij grondwaterwinning in Nederland, onderzoek naar alternatieve maatregelen (Nitrates by groundwater extraction in the Netherlands, investigations of alternative measures, in Dutch). Rapport 12 Institute for Land and Water Management Research, Wageningen.

Discussion Session 5

Questions from Mr. A. Armstrong to Mr. S.E. Olesen

I have three questions on the issue of ochre:

1) How did you map pyrite-rich areas which you then identified as potential ochre problem areas? This is something we have found to be very difficult, particularly the chemical measurement of pyrite content, which we have found to be very variable.

2) It appears that your concern about ochre arises from the pollution aspect, whereas for most of us here, the problem of ochre arises from the blocking of drainage systems. It appears that in the situations you describe, the iron remains in solution until it leaves the drainage system. It is our experience that this happens only when the drainage water is very acid. Is this the case in the situations you describe?

3) Have you tried using pine bark to absorb iron? This technique has been used by workers in Scotland with considerable success.

Answer of Mr. Olesen

The survey was done by profile examinations in all present and former wetland areas. On the average one profile covered about 75 ha. Soil samples were analyzed by a simple pH-decrease test and by direct analysis of the pyrite content. The areas were classified and mapped according to the percentage of pyrite-positive profiles. For land drainage a more detailed soil survey is carried out.

We recognized also problems of clogging of drainage systems. The problems in the watercourses appear when the ferrous iron concentration is very high and when the discharge from the drained areas constitutes a large portion of the stream flow.

We have investigated the ferrous iron retention in a small artificial pond filled with pine bark. The initial iron removal was almost 100%, but very soon problems of clogging of the bark by ochre arised.

Question from Mr. J. Mulqueen to Mr. E.A. Garwood

There was a larger N-loss from old pasture after drainage than from a reseeded pasture. When was the drainage and reseeding carried out in relation to the period of the measurements? Maybe the increase in loss to drainage water on old pasture was due to undisturbed cracks.

Answer of Mr. Garwood

313

The moles were originally drawn in at the end of August 1982, immediately before re-seeding. However, hydrographs from some plots showed that this moling, done when the soil was re-wetting, had not been very effective. Moling was therefore repeated in June 1983 and at present appears satisfactory. There is, however, evidence of less structure in the upper 20 cm soil after cultivation.

Questions from Mr. J. Bouma to Mr. H.C. Aslyng
1) You used the Addiscott flow model. How did you estimate the immobile and mobile fractions of water?
2) When studying the N-regime of clay soils we have observed very rapid leaching of N-fertilizer by bypass flow (rapid downward movement along cracks). Such losses are not related to denitrification. Did you observe identical phenomena?

Answer of Mr. Aslyng
1) We use the retention and the deficit of water in the soil profile. The distinction between mobile and immobile water is made at pF 3.3.
2) No. Our clay soils are relatively low in clay content (15-25%) and they are not cracking, so we did not observe the phenomena you mention.

Question from Mr. P.J.T. van Bakel to Mr. H.C. Aslyng
In the Netherlands there is a discussion about the economical feasibility of sprinkling irrigation on grassland, because from purely hydrological calculations the effects are too low to make sprinkling attractive. In my opinion, one important effect of sprinkling is the forced infiltration of applied fertilizer. Do you have data or ideas about this effect?

Answer of Mr. Aslyng
In case of water stress or deficit in the soil, irrigation is very important: a) to dissolve applied nitrogen and b) to utilize applied nitrogen. If no irrigation is applied the nitrogen is lost by leaching later on, so indeed irrigation may be advantageous in this respect.

Question from Mr. H.C. Aslyng to Mr. E.A. Garwood and Mr. P.E. Rijtema
You stated that drainage increased leaching of nitrogen from the root zone. According to my experience deep draining of light clay soils give possibilities for deeper rooting and less leaching of nitrogen. In autumn mineralized nitrogen to a great extent (under Danish precipita-

tion conditions) remains in the root zone and is available for the crop next year.

Answer of Mr. Garwood

In the heavy clay soils on our site the problem was largely surface water due to the very impermeable nature of the soil coupled with high rainfall. The lateral drains are 85 cm deep (typical of the drainage conditions of many UK clay soils) and it is unlikely that grass roots extend much beyond this depth in this soil. Excess winter rainfall is such that little of the previous years residue of nitrate remains into the following year - it is lost either through the drains or, on the undrained soils by denitrification.

Conclusions, recommendations and future activities

GENERAL

The objective of the workshop was to obtain a review of and to discuss methods that may lead to cost effective but environmentally acceptable techniques in agricultural water management.

Agricultural production of many soils in the Community is limited by soil water deficit and/or excess water. Improvement of water management may greatly influence types of crops to be grown, farm structure, costs of farm operation, income from and employment in agriculture.

In many cases the costs of improving water management structures are high. Additional problems may arise when available water resources are scarce and water management measures for agriculture have a negative influence on natural vegetation and fauna.

There is a growing need for methods to be used for the assessment of economic and environmental effects that can serve as a basis for developing efficient techniques in the improvement of agricultural water management. These techniques should be such that they give a reduction in costs and avoid negative effects on interests like nature, domestic and industrial water supply and so on.

The workshop was attended by 33 participants from 10 EC-member states and 1 representative of CEC. During the workshop 24 papers have been presented and discussed on the following topics:

1. Drainage and reclamation of soils with low permeability: 4 papers
2. Effects of drainage and/or irrigation on agriculture: 8 papers
3. Installation and maintenance of drainage systems: 5 papers
4. Regional and local water management systems: 3 papers
5. Effects of agriculture on its environment: 4 papers

On Thursday afternoon, 20 June, a visit took place to the Rhine-Waal river district. The excursion was organized by the Governmental Service for Land and Water Use and aimed to make acquaintance with the procedures, practice and results of integrated land consolidation schemes as applied now-

317

adays in the Netherlands. Delegates were unanimous in their opinion that the visit was most informative and instructive.

The participants of the workshop discussed in the final session of Friday 21 June, chaired by G.A. Oosterbaan, conclusions and recommendations, drafted by the chairman and secretary of each of the five sessions. Also the opinion of the participants about the framework of the future meetings was asked. It was unanimously held that the exchange of knowledge, experience, research methods and data between the EC-countries can be most effectively realized through:

1. workshops on specific subjects, and
2. exchange of scientists.

Scientific meetings on restricted research topics serve a higher priority than meetings with a wider scope. However, it was felt that the latter remain also necessary, but more occasional, to keep an eye upon the coherence of the research problems in the field of agricultural water management.

1. DRAINAGE AND RECLAMATION OF SOILS WITH LOW PERMEABILITY

Conclusions

1. Presence of stable macro-pores is essential for effective drainage in heavy soils. Morphological and physical techniques have been developed now to characterize these pores and these should be applied to different soils to better define drainage potential.
2. In the absence of natural macro-pores, fissures need to be created mechanically and stabilised through natural processes.
3. Where very close drain spacings are required making pipe drainage systems uneconomic, mole drains have potential for providing an economic drain.
4. Further progress on the improvement and extension of mole drainage practices based on soil and climatic studies alone, is likely to be very limited.

Recommendations

1. Investigate possibilities for changing mole plough design to improve performance and life of both leg fissures and mole channels. Identify specific implement designs for use under given soil and climatic conditions.

2. Certain standard assessment should be included in all moling studies, to assist with the interpretation and extension of the results. The assessments should include the following: texture, aggregate stability, predominant type of clay mineral, dry bulk density, macropore types and macropore continuity, moisture status at time of moling, time after moling before channel carries water, mole plough geometry, type of channel failure, rainfall/evapotranspiration data.

3. Assess feasibility and suitability of different mole channel collector drains for different conditions by gravel and pipe systems, mole drains, open channels.

4. Develop assessment methods for establishing whether artificial backfill materials are required to maintain the long term permeability of the soil in the drainage trench.

5. Use of soil survey information for extrapolating drainage data obtained for a given soil type should be explored, also among different countries in the EC. Contacts with the EC Working Group on Land Evaluation could be relevant in this context.

Future activities

The CEC is invited to organize an expert meeting on drainage of heavy clay soil in 1986. Ireland is disposed to host this meeting.

2. EFFECTS OF DRAINAGE AND/OR IRRIGATION ON AGRICULTURE

Conclusions

1. Considerable progress has been made and much is known about the effects of drainage and irrigation on growth of arable crops.

2. Grass as a crop has no value in itself until it is utilised by livestock whether by grazing or as winter feed. The utilisation of grass under different soil water regimes is little understood.

3. Growth of the grass crop, its quality and utilisation and the length of the growing season are strongly influenced by soil water regime, the season and by the conditions prevailing at previous grazings and harvestings. There is need for research to derive and elucidate relationships in this field.

4. Crop growth and production in relation to water management would appear to be a suitable subject for EC cross country collaboration to cover a variety of climatic and soil water conditions.

5. Models are increasingly being developed and tested. The results are prom-

ising but much work is still to be done. The detailed models are the most fundamental, but less detailed models require less input data. The objectives of application determine the degree of detail of models. At present models are contributing much to efficient planning of research and experimental work and will in the future be used more and more in advice, planning and management.

Recommendation

Activities should be initiated focussing on the development, application and validation of simulation models for cropping systems in relation to agricultural water management practices. They should pay particular attention to operational aspects, by defining an optimal degree of detail of the models and of data needed. Members of the Working Group on Land Evaluation, where simulation techniques are also used now, should take part in this activity.

Future activities

The CEC is invited to organise an expert meeting on development and application of simulation models for cropping systems in relation to water management in 1986. Belgium is disposed to host this meeting.

3. INSTALLATION AND MAINTENANCE OF DRAINAGE SYSTEMS

Conclusions

1. The problem of clogging of drains is complicated and progress of knowledge in this field is slow.
2. In several EC-member states, research is done on drain clogging problems: in Belgium, France and the Netherlands mainly.
3. Obviously, not all researchers in the member states publish their results in the commonly used languages. This is quite an unfortunate situation because it delays the progress of research efforts.
4. To enable relevant comparisons of drainage operating throughout the EC statistical analysis of field data should be developed, through similar methods.

Recommendations

1. Researchers should intensity their contacts on a more regular basis.
2. Research outcomes should be published in a language accessible to colleagues working in other countries, English by preference.

Future activities

The CEC is invited to organise an expert meeting in 1986 to discuss problems linked with mineral clogging mainly. France is disposed to host this meeting.

4. REGIONAL AND LOCAL WATER MANAGEMENT SYSTEMS

Conclusions

1. Water management practices differ not only from country to country, but even within every country, dependent on climate, hydrological and soil conditions and type of land use. Water management systems may be used for water discharge as well as for water supply or both.
2. The reports discussed during the meeting clearly show the wide variety of problems involved. Nevertheless the applied techniques in solving the problems may have a far wider scope of application.
3. The exchange of knowledge between the EC-countries therefore can be a valuable tool to obtain solutions for a wide variety of water management problems.

Recommendation

The research in the EC-countries should be directed to development of methods to use for manipulating water management on a regional scale.

Future activities

The diversity of discussed subjects did not enable to indicate specific research topics of common interest in this field for meetings on a short term.

5. EFFECTS OF AGRICULTURE ON ITS ENVIRONMENT

Conclusions

1. The presentation and discussions were restricted to the iron and nitrate problem imposed by changing water management. Other components may also have negative impacts on the environment in relation to changes in water management.
2. Water management affects environmental conditions, both in favourable and unfavourable directions. These effects have to be considered in the evaluation of improvement of agricultural water management. This evaluation can be subject to preset quality limits by interested parties that use groundwater and surface water resources.

3. The present knowledge is too limited to give a sound basis for imposing regulations, by policy making authorities, on agricultural water management in order to fulfill the preset goals.

Recommendations

1. It is recommended that studies on modelling of expected water quality due to the type of land use and changes in regional water management should be encouraged.
2. Research on basic processes, data collection and exchange of data between the EC-countries have to be strongly promoted.

Future activities

The workshop on agriculture and environment to be organized in Germany next year will meet the common demand of the participants for meetings to discuss the effects of agriculture on the environment.

List of participants

BELGIUM

Dr.ir. W. Dierickx   Rijksstation voor Landbouwtechniek
Van Gansberghelaan 115
B 9220 Merelbeke

Prof. J. Feyen   Faculteit voor Landbouwwetenschappen
K.U.L.
Kardinaal Mercierlaan 92
B 3030 Heverlee

Dr. D. Xanthoulis   Faculté des Sciences Agronomiques de l'Etat
Chaine de Génie Rural II
B 5800 Gembloux

DENMARK

Prof.dr. H.C. Aslyng   Hydrotechnical Laboratory
The Royal Veterinary and Agricultural
University
Bülowsvej 23
DK 1870 Copenhagen

Dr. S.E. Olesen   Hedeselskabet
Danish Land Development Service
Klostermarken 12
DK 8800 Viborg

Dr. P.C. Thomsen   Institute of Irrigation and Agricultural
Meteorology
Jyndevad
DK 6360 Tinglev

FEDERAL REPUBLIC OF GERMANY

Dr. B. Tischbein   Lehrstuhl für landwirtschaftlichen Wasserbau
und Kulturtechnik
Universität Bonn
Nussallee 1
Bonn

FRANCE

Dr. Ph. Lagacherie

INRA
Laboratoire de Science du Sol
CRAD de Montpellier
9 Place Viala
34060 Montpellier Cédex

Dr. B. Lesaffre

Ministère de l'Agriculture
CEMAGREF - Division Drainage
Boîte Postale 121
92164 Antony Cédex

GREECE

Dr. St. Chadgejannakis

Ministry of Agriculture
Land Reclamation Institute
Sindos 57400
Thessaloniki

Dr. S. Aggelides

Ministry of Agriculture
Soil Science Institute of Athens
1 Soph. Venizelou Street
14123 Lycovrissi, Attiki

IRELAND

Dr. L. Galvin

An Foras Taluntais
Kinsealy Research Centre
Malahide Road
Dublin - 5

Dr. J. Mulqueen

An Foras Taluntais
Creagh
Ballinrobe, Co. Mayo

ITALY

Dr. P. Bazoffi

Istituto Sperimentale per lo Studio e la Difesa
des Suolo
Piazza d'Aglezio 30
50121 Firenze

Dr. M.E. Venezian-Scarascia

Istituto Sperimentale Agronomico
Via Ulpiani 1
70100 Bari

LUXEMBOURG

Dr. Al. Puraye

Service Pédalogique
ASTA - Laboratoire de Contrôle et d'Essais
Boîte Postale 75
Av. Salentiny
9080 Ettelbruck

THE NETHERLANDS

Ir. P.J.T. van Bakel   Institute for Land and Water Management
Ir. J.J.B. Bronswijk   Research (ICW)
Ir. G.A. Oosterbaan   P.O. Box 35
Dr. P.E. Rijtema   6700 AA Wageningen
Ir. L.C.P.M. Stuyt
Dr. J. Wesseling
Dr. A.L.M. van Wijk

Ir. W.A. Dekkers   Research Station for Arable Farming and Field
Ir. J.J. Schröder   Production of Vegetables (PAGV)
Ir. H.H. Titulaer   P.O. Box 430
                              3200 AK Lelystad

Dr. J. Bouma   Soil Survey Institute (STIBOKA)
Ir. H. Wösten   P.O. Box 98
                              6700 AB Wageningen

Ir. P.J. Kusse   Governmental Service for Land and Water Use
                              Rijkskantorengebouw Westraven
                              P.O. Box 20021
                              3502 LA Utrecht

Ing. G.A. Ven   IJsselmeerpolders Development Authority
                              P.O. Box 600
                              3200 AP Lelystad

UNITED KINGDOM

Dr. A. Armstrong   MAFF
                              Field Drainage Experimental Unit
                              Anstey Hall
                              Meris Lane, Trumpington
                              Cambridge CB2 2LF

Dr. E.A. Garwood   Grassland Research Institute
                              North Wyke
                              Okehampton, Devon

Prof. G. Spoor   Dept. Agricultural Engineering
                              Silsoe College
                              Silsoe
                              Bedford MK 45 4DT

COMMISSION OF THE EUROPEAN COMMUNITIES

Dr. E. Culleton   Directory-General for Agriculture
                              Rue de la Loi 86
                              B 1040 Brussel
                              Belgium